Science
WORKBOOK 1

David Blaker, Anne Rowlands, Brent Smith

Science Workbook 1
1st Edition
David Blaker
Anne Rowlands
Brent Smith

Text designer: Cheryl Rowe
Cover designer: Cheryl Rowe
Production controller: Siew Han Ong

Any URLs contained in this publication were checked for currency during the production process. Note, however, that the publisher cannot vouch for the ongoing currency of URLs.

© 2011 Cengage Learning Australia Pty Limited

For product information and technology assistance,
in Australia call **1300 790 853**;
in New Zealand call **0800 449 725**

For permission to use material from this text or product, please email **aust.permissions@cengage.com**

National Library of New Zealand Cataloguing-in-Publication Data
National Library of New Zealand Cataloguing-in-Publication Data

Blaker, David (David Neville)
Science workbook. 1 / David Blaker, Anne Rowlands, Brent Smith.

ISBN 978-017021-465-0
1. Science—Problems, exercises, etc.—Juvenile literature. [1. Science—Problems, exercises, etc.]
I. Rowlands, Anne. II. Smith, Brent. III. Title.
507.6—dc 22

Cengage Learning Australia
Level 7, 80 Dorcas Street
South Melbourne, Victoria Australia 3205

Cengage Learning New Zealand
Unit 4B Rosedale Office Park
331 Rosedale Road, Albany, North Shore 0632, NZ

For learning solutions, visit **cengage.com.au**

Printed in China by 1010 Printing International Limited
9 10 11 12 13 25 24 23 22 21 20

The nature of science

What is science? It's our way of exploring and investigating the natural world, from atoms to animals, from earthquakes to energy. The process of discovery is not always straightforward. Science is always growing and changing, with a mixture of certain facts and uncertain information and ideas and theories. Why uncertain? In many situations – both living and non-living – we can't be sure because there is not enough evidence yet.

Science is built on three foundations: questions, ideas, and evidence. Without questions and curiosity there would be no progress. Without ideas and theories and imagination, science would just be a collection of facts. And as teams of scientists work at uncovering more evidence, ideas can become stronger.

Science is useful. It provides a platform for all technology and medical treatments, and also offers a basis for decisions on many social and environmental issues. It even helps you sort out sense from nonsense in aspects of everyday life.

Nature of Science is a description that sums up the main features that link the many parts of science. These features include questions, investigations, evidence, experiments, theoretical models, and practical uses. A *Nature of Science* strand runs through this book, and is a feature of many of the tasks.

The textbook SciencePLUS 1 contains detailed information, many practical experiments, more than 150 ideas for investigations and a student CD. Guidance is provided where units in this workbook link with units in SciencePLUS 1 (e.g. **SP1 Unit 1.2** = SciencePLUS 1 Unit 1.2).

Contents

Physical Science

Earth Science

Astronomy

Glossary   156

Periodic table of the elements   160

For Students

This science workbook is intended for use either at home, or in class. It can be used as a stand-alone resource, but is better used alongside the textbook *SciencePLUS 1* – which has much wider information. Each activity in this workbook is linked to a particular unit in the textbook. You can do almost all workbook tasks at home without a textbook, but in most cases the task will be easier if you have recently covered the linked unit in class time. You will find that the activities range from easy to difficult. Some may take you less than 10 minutes, some more than 20 minutes. Most tasks need short answers. The amount of space provided for each answer gives an idea of how much to write. Harder tasks have the word **CHALLENGE** in the margin.

It also helps to be clear about the type of answer needed in each situation. It may help you to visualise that each question and activity and skill belongs in one or other level of a three-storey building. Mostly, you will be building knowledge on the ground floor – but the higher floors will become more important as your knowledge increases. How can you know which floor you are supposed to be working on? Look for key words, especially verbs like: describe, explain, discuss.

Second floor: **Judging and imagining**
Look for any of these words:
Discuss, evaluate, imagine, judge, predict, design, plan, suggest, hypothesise.
EXCELLENCE

First floor: **Understanding and applying**
Look for any of these words:
Explain, compare, classify, sort, analyse, reason, summarise, arrange in sequence, apply a principle.
MERIT

Ground floor: **Knowledge and remembering**
Look for any of these words:
Describe, count, define, identify, list, match, name, observe, label, draw, select.
ACHIEVED

You will not find any research projects in this workbook but *SciencePLUS 1* has more than 150 ideas for investigations and projects, plus hundreds more questions, and 77 hands-on practical observations and experiments.

ISBN: 9780170214650

Start-up

This section contains information you need to start science.

- Some tasks deal with basic skills, safety measures, and uses of equipment.
- Other tasks deal more with the methods and 'nature' of science. In other words: how science works.

You could do all of the Start-up section at the beginning, or you could do some tasks – such as drawing graphs – only when you need to later on.

1 Safety first

| Date for completion: / / | Parent sig: _____ |
| | Teacher sig: _____ |

1   Basic safety rules need to be followed in any science room. Explain at least one reason for each of the following 12 rules. More than one reason applies in some cases. The first has been completed as an example.

A   Never taste laboratory chemicals, *because some of them are poisonous.*

B   Walk, don't run, *because...*

_____

_____

C   Wipe up any spills immediately, *because...*

_____

_____

D   When using a Bunsen burner always put a board under it, *because...*

_____

_____

E   School bags should either be left outside or at the front of the room, *because...*

_____

_____

F   Don't bring food or drink into the science room, *because...*

_____

_____

G   Never burn pencils or paper and especially don't burn plastic, *because...*

_____

_____

H   Keep your chemical area separate from your writing area, *because...*

_____

_____

I   Always keep glassware away from the edge of your table, *because...*

_____

_____

J   Report all breakages straight away, *because...*

_____

_____

K   Wear plastic safety glasses when working with boiling liquids, *because...*

_____

_____

L   Always replace the tops of chemical containers, *because...*

_____

_____

2 Identify all the things that are being done incorrectly in this cartoon of a science room. For each mistake, neatly rule a line to the areas above or below the cartoon. At the end of each line, write a few words to explain what is being done wrongly.

ISBN: 9780170214650

2 Getting into gear

Date for completion: / / Parent sig: _____ Teacher sig: _____

SP1 Unit 1.2

1 Next to each drawing, write the correct name of the item of equipment. This list provides eight of the 12 names: *beaker, conical flask, measuring cylinder, tripod, funnel, gauze, Bunsen burner, spatula.*

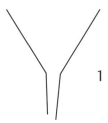 1

2

3

4

5

 6

7

8

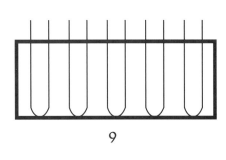 9

10

11

12

1 _____   2 _____

3 _____   4 _____

5 _____   6 _____

7 _____   8 _____

9 _____   10 _____

11 _____   12 _____

2 Complete this table by writing the number and name of each equipment item next to the description of its use.

Number	Use	Name
	For chemical reactions with larger amounts of liquid	
	For chemical reactions with small amounts of liquid	
	To produce a flame and heat things	
	To spread a flame evenly	
	To hold a large amount of liquid	
	A stand to put things on while they are being heated	
	For accurately measuring the volume of a liquid	
	For holding filter paper when filtering	
	To scoop up dry powder, but not for stirring	

3 Curious?

Date for completion:	/ /	Parent sig: _____
		Teacher sig: _____

SP1 Unit 1.3

Science begins with curiosity. Its main goal is to find as much as possible about the natural world and how it works. Another feature of science: it is **empirical**.

Next to each of the three photos below, make a list of observations on details from that particular photo. At the end of your list, write one or two questions about features of the photo that are interesting or puzzling to you.

> **Empirical** (*adjective*): any knowledge based on experience, evidence or testing; and not based on ideas alone. This is a key feature of how science works.

_____
_____
_____
_____

_____
_____
_____
_____

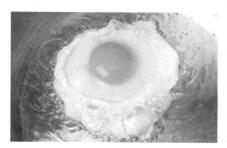

_____
_____
_____
_____

ISBN: 9780170214650

4 Evidence, prediction, proof

Date for completion: / /

Parent sig: _____
Teacher sig: _____

Science depends on evidence. Good evidence plus clear thinking is the scientific way to test ideas. Without good evidence, ideas become wobbly. One problem: evidence can get mixed up with predictions and opinions.

> **Evidence:** facts or signs showing that something exists or is true.
> **Proof:** a collection of evidence that is 100% convincing, or almost 100%.
> **Inference:** something that you think is true, or possibly true, based on evidence. If you are very certain, an inference could be called a **conclusion.**
> **Prediction:** a statement about what you think is going to happen. Solid evidence leads to good predictions; weak evidence leads to weak predictions.

Below is a list of statements, arranged in pairs: AB, CD, EF etc. Read each statement carefully and decide whether it is (mainly) **evidence**, or **inference**, or **prediction**, or **proof**, or **conclusion**. Write ONE of these words at the end of each statement. Word meanings are given in the box above.

A My cat doesn't like milk. _____ Evidence_____

B *There is probably something in cow's milk that is bad for him.* _____ Inference_____

C The forecast is for rain on Friday. _____

D *That probably means the weather will be fine by Sunday.* _____

E My brother ate a whole big slab of chocolate yesterday. _____

F *So he will probably get an outbreak of pimples soon.* _____

G Cicadas are very noisy the summer. _____

H *My gran says this probably means we will get a cold winter.* _____

I Auckland has about 50 extinct volcanoes. _____

J *Auckland is likely to have volcanic eruptions in the future.* _____

K The sedimentary rocks around here are older than the volcanic rocks. _____

L *Because the volcanic rocks all lie on top of sedimentary rocks.* _____

M Whenever you pump up a tyre, the pump gets hot. _____

N *The heat energy has come from kinetic (mechanical) energy.* _____

O My dog barks and howls when she is left alone. _____

P *She will probably bark less if I take her on a long walk every day.* _____

Q Babies who get lots of hugs turn out to be happier adults. _____

R *Early experience affects personality.* _____

S On average, smokers die about 10 years younger than non-smokers. _____

T *Smoking is bad for health.* _____

5 **Using a Bunsen burner**

Date for completion: / /

Parent sig: _____
Teacher sig: _____

SP1 Unit 1.4

Start-up

1 Complete the following. Some words may be used more than once: *blue, yellow, closed, open, air, gas, barrel, cooler, visible, collar.*

To light a Bunsen burner, start with the air hole _____. Next, turn

the _____ tap on, then immediately put a lighted match above the

_____ This will give a _____ colour flame that is

not very hot. To get a hotter flame, open the _____ to increase the

_____ supply. The resulting flame colour is then _____.

If you are not going to be using the Bunsen for more than a few seconds, best close the

_____ so that the flame is coloured _____. Two

reasons why this is safer: the flame is _____, and _____

2 Label parts A to G in the drawing.

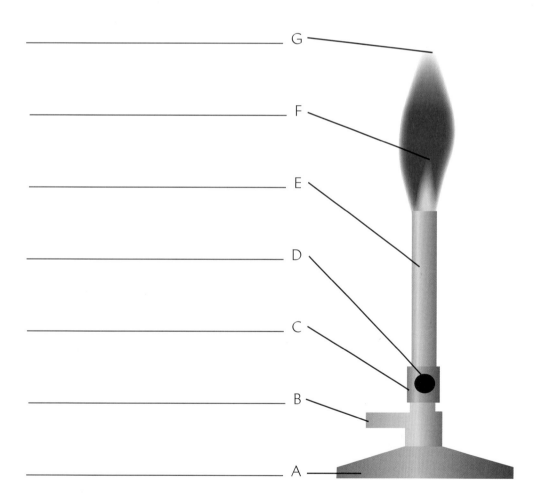

_____ G

_____ F

_____ E

_____ D

_____ C

_____ B

_____ A

ISBN: 9780170214650

6 Experiment words

Date for completion: / /   Parent sig: _____   Teacher sig: _____

SP1 Unit 1.3/1.4

Use the following clues to complete the puzzle. All 12 words relate to experiments, or to the nature of science in general. To help you, some letters have been provided.

1 A general statement that is based on result.
2 Best if your results are written in one of these.
3 An idea or suggestion that can tested.
4 Means the same as 'trial' or 'try-out'.
5 Can mean an opinion, or can mean a big idea.
6 Anything that could affect your results.
7 Write it down straight away!
8 Means recorded information, words, drawings, measurements.
9 What you hope to find out.
10 Science usually starts with these.
11 Best if your results include these.
12 Anything you see, hear, or measure.

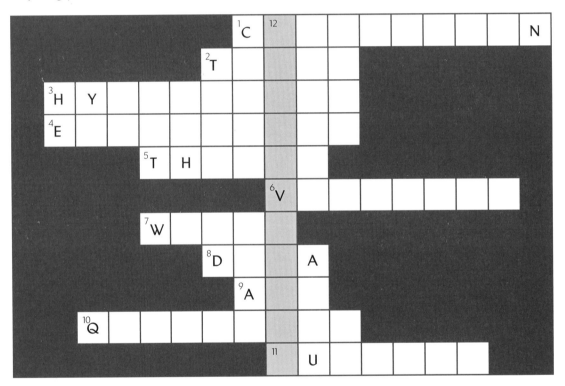

7 Hypo what?

Date for completion: / /   Parent sig: _____   Teacher sig: _____

SP1 Unit 1.4

There are plenty of new ideas in science, but not all of them turn out to be correct, so there are ways of sorting out correct ideas from wrong ones. Testing ideas in an organised way is an important part of scientific method. This process of idea-testing is not limited to classrooms: it is part of everyday life.

> **Hypothesis:** an idea or a prediction that can be tested. In many situations, a hypothesis can be built around the words 'If..., then...'.

Example: Farmer Joe has heard of the idea that music helps cows produce more milk. Some people find the idea reasonable, and some think it is absurd – but argument alone will not sort this out. On the next page, you are to plan an organised trial to test Joe's idea.

Hypothesis

'If I play music in the cowshed,
then my cows may produce more milk.'

Trial (experiment)

To test this idea, Joe would need to do at least two trials. The reason for doing two trials is to give a basis for comparing.

Trial 1 (describe it)

_____

_____

Trial 2 (describe it)

_____

_____

Keep it fair! List at least four or more **variables** that Joe should try to keep the same for his two trials, in order to make the comparison as fair as possible.

_____

_____

_____

_____

_____

_____

Results

Suggest what Joe should measure in order to get reliable results. (The 'thing' being measured to get results is also known as the **dependent variable** because it may depend on which variable the planner has changed.)

_____

_____

_____

Repeats

Suggest why any conclusion is more reliable if trials are repeated a few times.

_____

_____

_____

CHALLENGE

Flexible or fixed?

Science is not just a collection of facts; it is more a process of discovery. It is about asking questions, coming up with ideas, testing predictions. Also, not everything in science is definite. Ideas can change over time; some are almost completely certain, some less certain. Scientific methods help find which ideas are backed by good evidence, and which are not.

ISBN: 9780170214650

Start-up

1   Complete the labelling of these two photographs using these labels: *light source, stage, clip, eyepiece lens, objective lenses, coarse focus control, fine focus control, base, stereo microscope.*

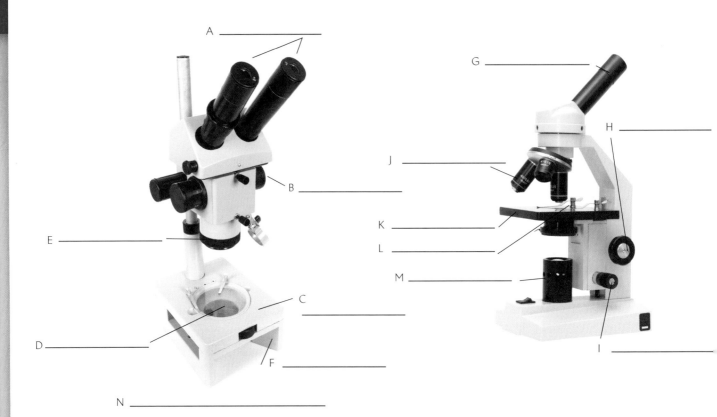

A _____

G _____

H _____

J _____

B _____

K _____

E _____

L _____

M _____

C _____

D _____

I _____

F _____

N _____

2   Supply the missing words. Here are four to help you get started: *top, low, stage, underneath.*

When using any kind of microscope, always begin with the objective lens set to

_____ power. For solid objects, best use _____ lighting,

and for transparent objects best use _____ lighting. Place the object

you want to look at on a _____ dish or a _____,

then put it in the centre of the _____. Next: focus gently.

If a microscope objective lens is 4x and the eyepiece is 10x, the combined magnification

is _____. One important 'don't' for microscope use is: don't

_____

_____

9 ▶ Microscope sizes

Date for completion: / /　Parent sig: _____　Teacher sig: _____

SP1 Unit 2.1

Measure the body length of microscopic life A, B, C and D in millimetres (mm) using the scale underneath. Then convert these to micro-metres (= 1/1000 mm).

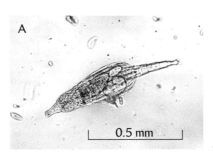

A

0.5 mm

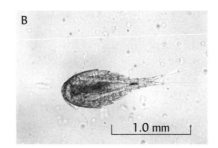

B

1.0 mm

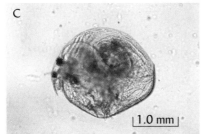

C

1.0 mm

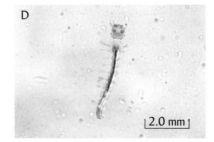

D

2.0 mm

A　Actual length is about _____ mm, which is _____ micro-metres.

B　Actual length is about _____ mm, which is _____ micro-metres.

C　Actual length is about _____ mm, which is _____ micro-metres.

D　Actual length is about _____ mm, which is _____ micro-metres.

10 ▶ Measuring and graphing

Date for completion: / /　Parent sig: _____　Teacher sig: _____

SP1 Unit 4.3

A graph is a kind of picture that clearly shows patterns in a collection of numbers. There are different ways of drawing graphs. In the following tasks use the TADPL guidelines: Title, Axes, Divide, Plot, Line.

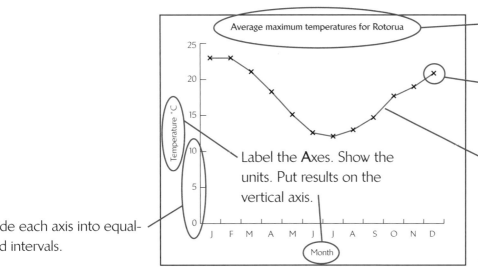

Give your graph a **T**itle.

Plot the data. Each pair of numbers is marked with a small neat cross.

Label the **A**xes. Show the units. Put results on the vertical axis.

Draw a **L**ine of 'best fit', whether curved or straight.

Divide each axis into equal-sized intervals.

ISBN: 9780170214650

1 Start with one page of newspaper, and measure the length of the longest side in millimetres (mm). Write this measurement in the table below. Now fold the paper in half and measure the 'new' longest side. Repeat this process for 2, 3, 4 and 5 folds.

Number of folds	Length of the longest side (mm)
0	
1	
2	
3	
4	
5	

2 Use the data in your table to make a graph that shows the situation. Use the TADPL guidelines.

3 Extrapolate the line of your graph, using it to predict the length of the longest side after 9 folds.

Answer:_____ mm after nine folds.

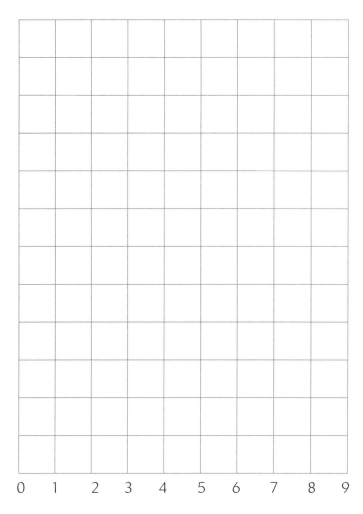

Number of folds

ISBN: 9780170214650

Date for completion: / /

Parent sig: _____
Teacher sig: _____

SP1 Unit 4.3

Start-up

For each of the following, write down the scale measurement.

1

A _____

B _____

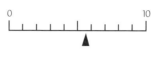

C _____

D _____

E _____

F _____

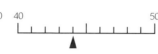

G _____

H _____

2

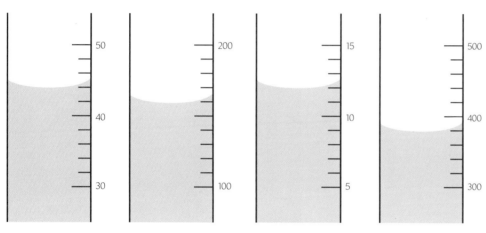

A _____ mL   B _____ mL   C _____ mL   D _____ mL

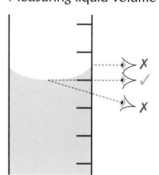

Measuring liquid volume

When you are measuring the volume of any liquid, look at the level centre of the meniscus surface, seen side-on.

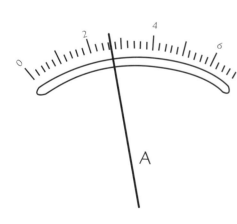

E _____ amps

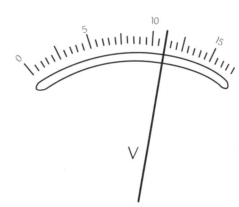

F _____ volts

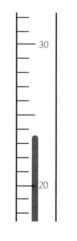

G _____ °C

ISBN: 9780170214650

12 Research guide

Date for completion: / / Parent sig: _____ Teacher sig: _____

SP1 Generic

This page offers help with research projects.

The first step: choose wisely!

✓ *Do choose a topic that interests you.*

✓ *Do choose a topic where you can easily get information.* Avoid topics where information is hard to find. It's a bad idea to change topics very late!

Whichever topic you choose, it's best to have an overall plan or 'road map'.

A good plan to follow is **Find-Select-Report: FSR.**

> **FIND**
> *Find* a range of information on the topic from books, magazines, internet etc.

> **SELECT**
> *Select* useful and reliable information from three or more different sources.

> **REPORT**
> *Report* and arrange your selected information in a clear and interesting way.

As well as using the FSR plan, make sure you start with facts on these five points.

1 Working in groups, or individually?

Answer: _____

2 School time only, or mainly homework?

Answer: _____

3 What day must it be finished by?

Answer: _____

4 How long should my report be?

Answer: _____

5 How will it be marked? Success criteria?

Answer: _____

FIND

The main places to find information:

- Books
- Interviews with experts
- Newspapers and magazines
- The internet

Books? A library is the best place to start. The science section is between classification numbers 500 and 599. If you choose a specialised topic, books may be hard to find.

Newspapers and magazines? Can be good, but it may be difficult to find more than one on your topic.

Talk with an expert? There are plenty of local experts around on fishing, machinery, cooking, cars, farming, building, manufacturing, astronomy, animals, medicine etc.

Internet? Excellent if you know exactly what you need, but searching can waste hours of your time. Most information is useless for your purposes. Select with care. This table shows that a search can be narrowed 99% by adding a few carefully-chosen words.

Search words used	Sites found
Elephants	13 million
Elephants ivory	1 million
Elephants ivory trade ban	214,000
Elephants ivory trade ban success failure	102,000

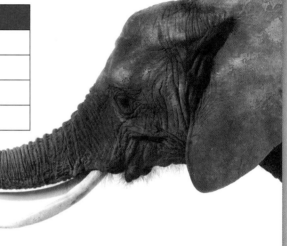

SELECT

✓ Select only what best suits your purposes.
✓ Decide the order in which information will be arranged.
✓ About half must be in your own words, not all copied/pasted.
✓ Ideally you need to use a total of three to seven helpful books, websites, magazine articles, people. Keep a list. Pick out the best information.

REPORT

Your teacher will give guidelines on what is needed in the final report. If you are working in a group, decide who will be responsible for which sections and which illustrations. Four possible ways of displaying (reporting) any assignment:

> Make sure that you include a **reference list** of the information sources you used – book titles, magazine titles and dates, website addresses.

✓ A4 sized pages.
✓ Poster: big, bright, eye-catching, not too many words.
✓ Creating a computer display using PowerPoint or similar software.
✓ A talk to the class, perhaps backed up by your poster or PowerPoint.

Bad science
Especially with websites and blogs, watch out for:

• advertising in disguise
• general statements based on just one example
• opinions stated as facts.

If you find even one of these features, delete the website from your list.

ISBN: 9780170214650

Life Science

1 Plant and animal cells

Date for completion: / /   Parent sig: _____ Teacher sig: _____

1 Complete the labelling of this plant cell diagram for parts 1 to 8.

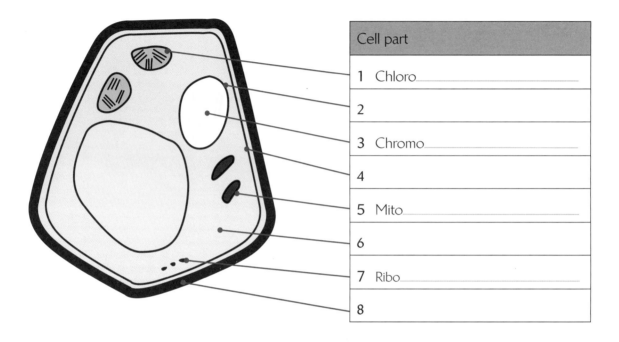

Cell part
1 Chloro_____
2
3 Chromo_____
4
5 Mito_____
6
7 Ribo_____
8

2 State the function of these cell features. Two have been done for you.

Cell part	Function
1 Chloro_____	
3	Stores information, controls cell chemistry
4	Regulates the entry and exit of substances
5 Mito_____	
7 Ribo_____	

3 Identify three features of the diagram that are unique to plants.

_____

_____

_____

Date for completion: / /

Parent sig: _____

Teacher sig: _____

Life Science

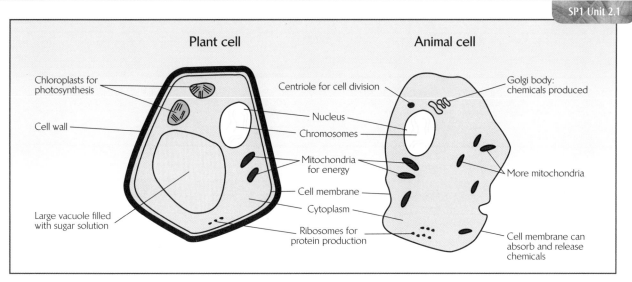

1 Write letters A to H in the appropriate areas on the Venn diagram below, to represent cell structures that exist in animal cells only, plant cells only, or both.

A cell wall   B cell membrane   C chloroplasts
D cytoplasm   E mitochondria   F large vacuole
G nucleus with DNA   H ribosomes

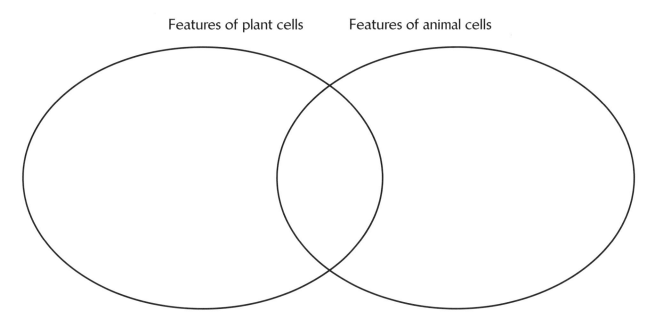

Features of plant cells   Features of animal cells

It helps to know where words came from, and how they are put together.
Many science words come from the Greek and Latin languages.
Examples: *chloro* = green; *chromo* = colour; *soma* = body; *cyto* = cell; *mito* = thread;
chondrion = little grain; *photo* = light; *synthesis* = put together; *plastis* = to mould or form.

2 Using the above information, explain the literal meanings of:

chromosome _____

chloroplast _____

mitochondrion _____

photosynthesis _____

ISBN: 9780170214650

Date for completion: / /   Parent sig: _____ Teacher sig: _____

SP1 Unit 2.1

Measure the length of the largest cell in each of A, B, C and D as they appear on this page. Write each measurement in the table. Now use the enlargement factor to calculate the actual length of each cell.

A
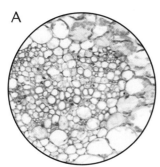
Cross-section of a plant vein

B
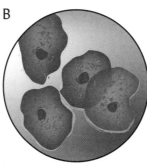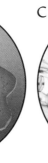
Epithelial cells from the lining of your mouth

C
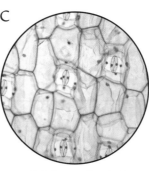
Cells from a plant leaf

D
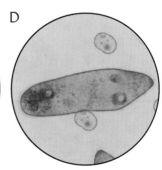
Paramecium

Cell	Greatest length in the photograph (mm)	Enlargement factor	Actual cell length in mm	Actual cell length in micro-metres
Example	10 mm	40 x	0.25 mm	250
A		40 x		
B		100 x		
C		200 x		
D		100 x		

Date for completion: / /   Parent sig: _____ Teacher sig: _____

SP1 Unit 2.2

1 The following five sentences have been scrambled so that the three main parts do not match. Rewrite all the sentences in the spaces below, so that each makes a correct statement of fact.

	Subject	Verb or verb phrase	Object
A	A tissue	is	the simplest level of life organisation
B	An organ	consists of	many cells with the same function
C	A kidney	consists of	signals from place to place
D	Nerve cells	are	an example of an organ
E	Cells	carry	many tissues acting together

Correct statements:

A   A tissue _____

B   An organ _____

C   A kidney _____

D   Nerve cells _____

E   Cells _____

Life Science

2 Place each of the following in its correct box: *muscle; brain; pancreas; nerve cell; tongue; heart; eye; brain + nerves + sense organs; blood; cartilage.*

CELL	TISSUE	ORGAN	SYSTEM

5 Before birth

Date for completion: / /

Parent sig: _____

Teacher sig: _____

The following eight sentences have been scrambled so that the three main parts do not match. Rewrite all the sentences in the spaces below, so that each makes a correct statement of fact.

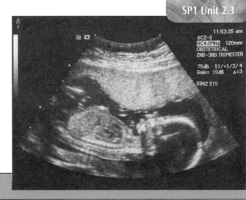

SP1 Unit 2.3

	Subject	Verb or verb phrase	Object
1	Amniotic fluids	begin at	the lungs before birth.
2	The umbilical cord	connects	if the baby is in a dangerous position.
3	Heartbeats	is where	about four weeks after fertilisation.
4	The uterus	begin	the baby's circulation with the placenta.
5	Caesarean section	is where	the foetus develops.
6	Breathing movements	may be necessary	mother and foetus blood meet but do not mix.
7	Blood	surround	the developing foetus.
8	The placenta	mainly bypasses	about 12 weeks.

Correct statements:

1 Amniotic fluids _____

2 The umbilical cord _____

3 Heartbeats _____

4 The uterus _____

5 Caesarean section _____

6 Breathing movements _____

7 Blood _____

8 The placenta _____

ISBN: 9780170214650

6 Boy or girl?

Date for completion: / /　Parent sig: ____________ Teacher sig: ____________

SP1 Unit 2.4

1 Every body cell has 46 chromosomes (23 pairs), but the diagram below shows only the sex chromosomes, known as X and Y. The other 44 non-sex chromosomes are not shown. Gametes (sex cells) have only one chromosome of each pair. Complete the writing in the spaces provided. Choose from: *X, Y, XX, XY, boy, girl.*

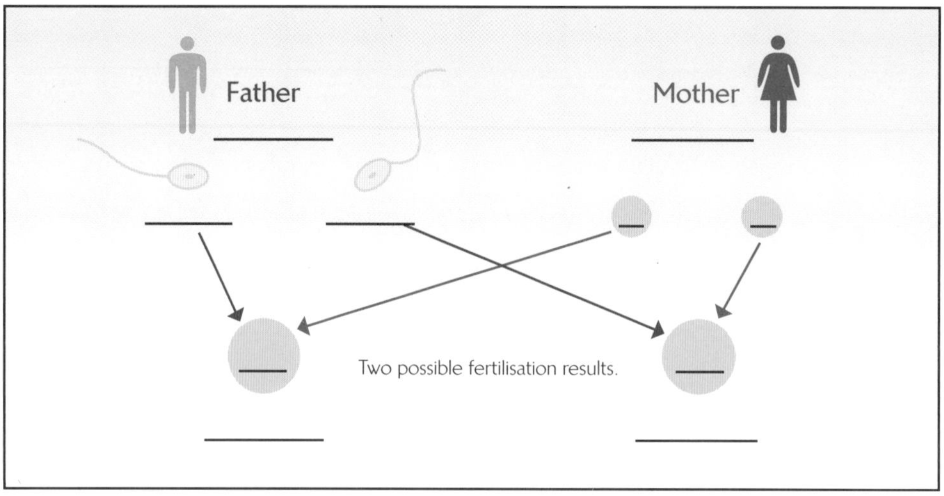
Father
Mother
Two possible fertilisation results.

2 Complete the sentences below by writing in items chosen from this list: *XX, XY, X, Y, 25%, 50%, 75%, will, will not.*

Because half of all sperm cells carry a ____________(1) chromosome, there is theoretically a ____________(2) chance of a boy being born from each fertilisation. In reality, the proportion of boys and girls generally is about ____________(3) to ____________(4). If a couple's first child is a girl, this ____________(5) increase the chances of a boy at their next pregnancy.

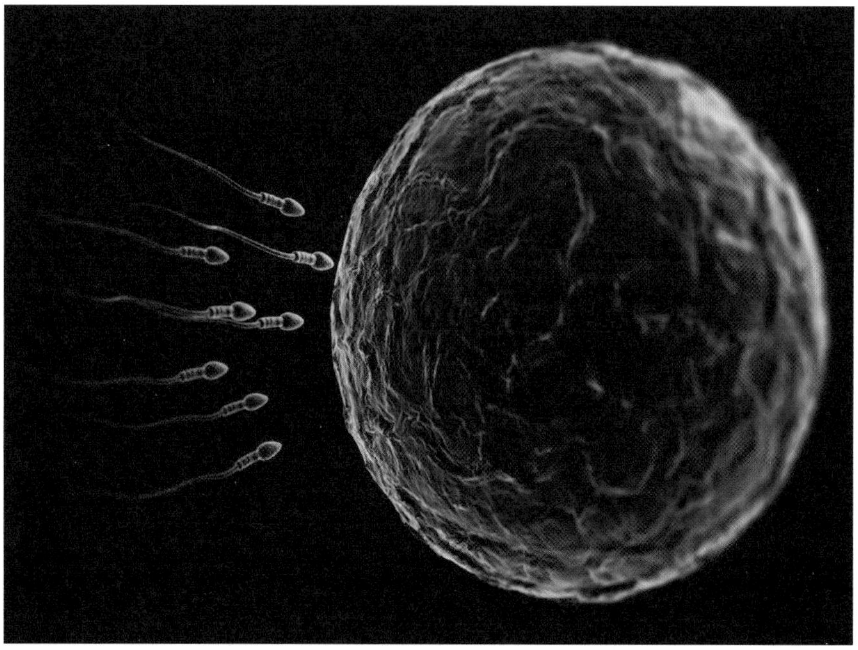

ISBN: 9780170214650

Twins and triplets

Date for completion: / /

Parent sig: _____
Teacher sig: _____

SP1 Unit 2.5

1   Complete the sentences below by writing in words/numbers
    chosen from this list: *25%, 50%, 100, 500, identical, non-identical.*

Normally one egg is released in each menstrual cycle, but occasionally

two are released. If both eggs are fertilised, this will produce two

babies with slightly different genes: _____(1) twins.

Twins like these are born on average in about one pregnancy in every

_____(2).   These twins could be two brothers

(_____(3) of cases), two sisters (_____(4)

of cases), or brother and sister (_____(5) of cases).

Much less common are _____(6) twins. Here, one

egg is fertilised by one sperm, and the developing embryo divides

completely in two, and the two parts go on to produce twins who

are genetically _____ (7). This happens in about one in

_____(8) pregnancies.

Life Science

2   Triplets can result from three eggs being produced at once with
    three separate fertilisations and three non-identical children born
    at the same time. Identical triplets are much less common.   The
    diagram shows the chance of a fertilised egg splitting in two,
    and one of these in two again. The chance of this happening is
    theoretically about 1 in 500 x 500 = 1 in 250,000.

    Using this information, calculate the theoretical chances of:

    Identical quadruplets

    _____

    _____

    _____

    Triplets, two identical, one not

    _____

    _____

    Non-identical triplets

    _____

    _____

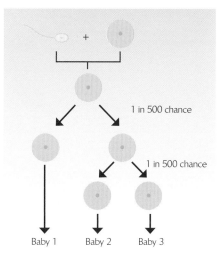

1 in 500 chance

1 in 500 chance

Baby 1   Baby 2   Baby 3

ISBN: 9780170214650

Life Science

Life is dynamic; it is not a list. However, the mnemonic 'MRSGREN' is sometimes used to list seven features of living things. Some of these seven features also exist in non-living situations.

Complete the table below. Most boxes can be completed by writing yes or no.

	Stands for:	Candle flame	River	Shark	Oyster	Rabbit	Tree
M							
R							
S							
G							
R							
E							
N							
	Score out of 7:						
	Alive or not ?						

1 There are perhaps 5 million or more different kinds of living species in the world, but so far less than 2 million species have been named. They are grouped in five main categories called 'Kingdoms'. Match each of these Kingdom names to the description in the table: *Fungi, Plants, Protists, Bacteria, Animals.*

Description	Name of Kingdom
Eat food produced by other living things.	
Make their own food by photosynthesis.	
Most kinds absorb food from decay. No photosynthesis.	
One cell, have a nucleus. Example: *Paramecium*.	
One cell, very simple, have no nucleus.	

2 For convenience, animals are classified into groups and subgroups. The hierarchy of classification is shown in in this diagram. Make up your own mnemonic to help you remember the sequence: KPCOFGS.

_____
_____
_____
_____
_____
_____
_____
_____
_____
_____
_____

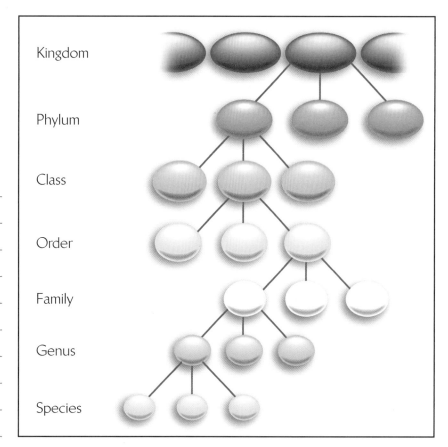

Kingdom

Phylum

Class

Order

Family

Genus

Species

3 The numbers of known species in some animal groups: 1 million insects; 100,000 molluscs; 40,000 crustacea; 24,000 fish; 9,000 birds; 4,500 mammals. Create a labelled bar graph representing these numbers.

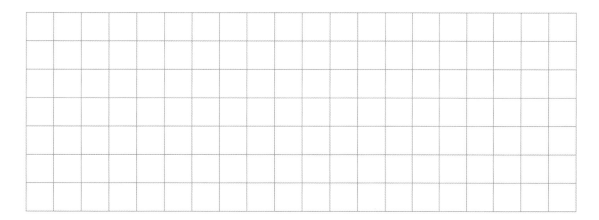

CHALLENGE

4 Animals are usually grouped according to their physical features. For example: only mammals have fur, only birds have feathers. Some books organise animals by name, alphabetically. Suggest one advantage and two disadvantages of an A–Z system for animal names.

_____
_____
_____
_____
_____

ISBN: 9780170214650

Life Science

Here is a list of 30 animals. Write the name of each inside its correct group below.

frog, snake, albatross, rabbit, beetle, spider, centipede, snail, earthworm, shrimp, eagle, dog, crocodile, penguin, crayfish, bee, scorpion, butterfly, dolphin, tuna, monkey, octopus, kina, parrot, oyster, horse, seahorse, gecko, mosquito, panda.

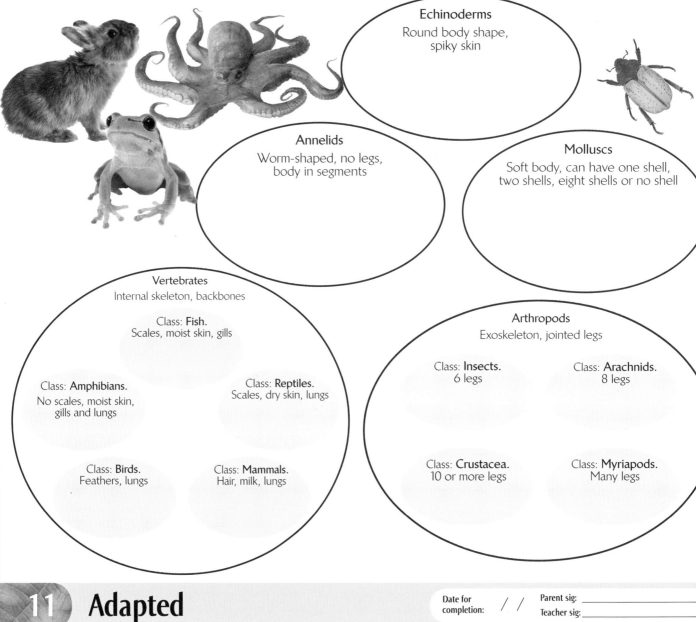

Echinoderms
Round body shape, spiky skin

Annelids
Worm-shaped, no legs, body in segments

Molluscs
Soft body, can have one shell, two shells, eight shells or no shell

Vertebrates
Internal skeleton, backbones

Class: **Fish.**
Scales, moist skin, gills

Class: **Amphibians.**
No scales, moist skin, gills and lungs

Class: **Reptiles.**
Scales, dry skin, lungs

Class: **Birds.**
Feathers, lungs

Class: **Mammals.**
Hair, milk, lungs

Arthropods
Exoskeleton, jointed legs

Class: **Insects.**
6 legs

Class: **Arachnids.**
8 legs

Class: **Crustacea.**
10 or more legs

Class: **Myriapods.**
Many legs

11 Adapted

| Date for completion: | / / | Parent sig: _____ |
| | | Teacher sig: _____ |

SP1 Unit 2.8

1 Complete the sentences below by writing in words chosen from this list: *behaviour, physiology, habitat, adapted, structure, adaptable.*

_____ means 'genetically fitted for a particular environment and way of life'.

Features that help an animal survive could be its _____ (physical body features), its _____ (what it does), and it's _____ (body chemistry).

2 Several features make bluefin tuna
   well-adapted as ocean predators.
   They are able to swim long distances
   at high speed, and they hunt
   swimming prey by sight not smell.
   Explain the 'survival value' of each of
   the following features.

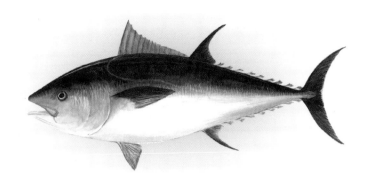

A Large eyes

_____

_____

_____

B Streamlined shape

_____

_____

_____

C Large gills

_____

_____

_____

D Pointed snout

_____

_____

_____

E Very muscular body

_____

_____

_____

F Fins can be folded away into grooves

_____

_____

_____

G Dark blue colour above, silvery below

_____

_____

_____

ISBN: 9780170214650

12 Kinds of adaptation

| Date for completion: | / / | Parent sig: _____ |
| | | Teacher sig: _____ |

For each of the following examples decide whether it is a **behavioural**, **physiological** or **structural adaptation** and explain the *function* of the adaptation.

> **Function**: its particular usefulness for survival.
> **Adaptation**: any feature that increases the chances of surviving, living longer, and breeding successfully.

Example	Type of adaptation	Function
A hedgehog can roll up into a ball.		
Penguin colour (black above, white below while swimming).		
Viper snakes inject venom from their fangs.		
Kiwi birds are active at night.		
Poinsettia plants are poisonous if eaten.		
Muttonbirds (titi) migrate from New Zealand to Alaska & back, every year.		

Date for completion: / /

Parent sig: _____

Teacher sig: _____

Design your own make-believe animal. First choose its habitat (e.g. desert, grassland, depths of the ocean, Antarctica, woodlands etc). Now design your animal adapted for that environment. Draw it in the space below. In the table list two or three features of each type of adaptation, and explain *why* it has this adaptation.

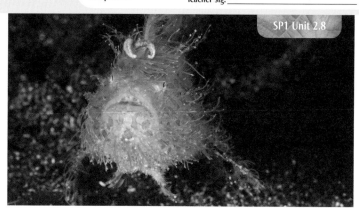

SP1 Unit 2.8

Life Science

Animal's name: _____

Type of animal: _____

Habitat where my animal lives: _____

Food my animal eats: _____

Predators my animal might have: _____

Adaptations:

Behavioural	Structural	Physiological

ISBN: 9780170214650

Life Science

Read the following information, and answer the questions below.

Mammals vary enormously in size, from two-gram pygmy shrews up to 100-tonne whales. Within this variety, only one feature is true of all mammals and is also unique to mammals: the females produce milk. In fact the word mammal is based on mamma, meaning mother. Hair is also unique to mammals, although some, like dolphins and most whales, are hairless. Most mammals have live birth, although two species lay eggs: duck-billed platypus and spiny anteater. Features like live-birth and air-breathing and having a warm body are however not unique - many fishes and reptiles have live birth, and birds have warm bodies. The 4,475 known species of living mammals are divided into 21 subgroups, also known as orders. The most successful mammal order is the rodents, with over 1,700 species, including rats, mice, beavers and squirrels. At the other extreme there are only two living species in the elephant order, although there were more species in the past. Monkeys, apes and humans are grouped in the primate order, all of which have grasping hands and forward-facing eyes. Examples of other mammal orders: seals, bats, sloths.

Another well-known order is the carnivores: dogs, cats, bears, hyenas, weasels, etc. Subgroups of an order, as in this example, are known as families.

1 How many living species of mammals are known?

_____

2 What is the name and size of the smallest mammal?

_____

3 Identify two features that are unique to mammals.

_____

4 Identify two other features that occur in almost all mammals, but also in some other living things.

_____

5 How many mammal orders are there?

_____

6 Which is the most successful order of mammals?

_____

7 Explain in what way this order is considered to be successful.

_____

_____

8 In which order are humans placed in? Explain why.

_____

_____

_____

9 Which order and family are tigers placed in? Order_____, family_____

_____

10 Which order and family are wolves placed in? Order_____, family_____

Date for completion: / /   Parent sig: _____
     Teacher sig: _____

SP1 Unit 2.11

Complete the crossword.

ACROSS
3 Chemical process that releases energy from sugar
5 It's red and it carries oxygen
6 Inflammation of these causes bronchitis
7 The tiny endings where gas exchange happens
8 Separates your chest cavity and your abdomen
10 The small air passages of the bronchi
13 Diffusion of oxygen in, carbon dioxide out
14 Muscles between your ribs

DOWN
1 Can cause your lungs to fill up with fluid
2 Waste product of cell respiration
3 Rubbery stuff, softer than bone
4 Can cause you to lose your voice
9 Also known as your main windpipe
10 Keeps on going, fortunately
11 Makes up almost 21% of the air
12 Your voice box

ISBN: 9780170214650

16 Gas exchange

Date for completion: / /   Parent sig: _____   Teacher sig: _____

SP1 Unit 2.11

The term 'gas exchange' includes the processes that move oxygen from air to cells, and carbon dioxide in the opposite direction. The following 10 steps in the overall process of gas exchange are listed out of order. Arrange them in the correct sequence in the boxes below, starting with A.

A the diaphragm contracts and flattens
B cell respiration: glucose and oxygen react to release energy
C as a result, fresh air is pulled into the lungs
D CO_2 is loaded from muscle cells into the blood
E oxygen diffuses from the alveoli into the blood
F CO_2 moves from the blood into the alveoli
G haemoglobin unloads its oxygen, which then moves into the cells
H haemoglobin inside red blood cells picks up oxygen
I blood travels from muscles to heart and lungs
J blood travels from the lung surfaces, eventually reaching your muscles

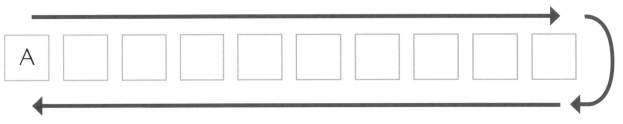

A

The 'steps' are cyclical and endlessly repeated.

17 Lungs and blood

Date for completion: / /   Parent sig: _____   Teacher sig: _____

SP1 Unit 2.11/2.12

1 Complete the labelling of these drawings, using the following words: *intercostal muscles, trachea, heart, bronchiole, bronchi, alveolus, air, diaphragm.*

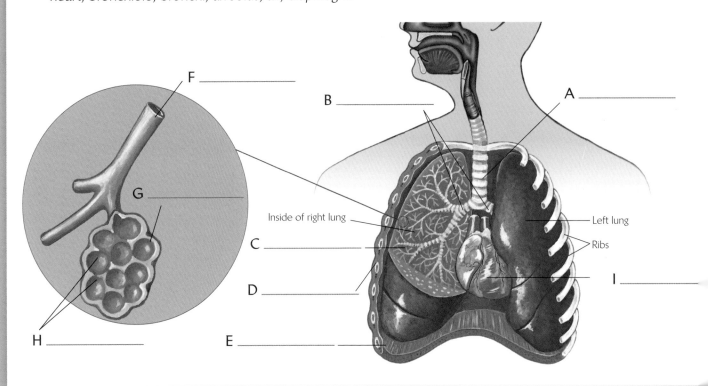

F _____
B _____
A _____
G _____
Inside of right lung
Left lung
C _____
Ribs
D _____
I _____
H _____
E _____

It's easy to mix up these three terms, but they have different meanings.
Breathing: movements of your muscles causing air to move in and out of lungs.
Cell respiration: the energy-releasing chemical reactions inside active cells.
Gas exchange: the movement of oxygen inwards, and CO_2 outwards.

2 At the lung surfaces, oxygen is loaded into the blood. Most of this oxygen is unloaded less than one minute later and used in active cells – especially brain and muscles. Oxygen's real work happens inside the cells, where it breaks down glucose. The breakdown of glucose is known as **cell respiration**, and it releases energy. Cell respiration can be summed up in this word equation:

oxygen + _____ → _____ + _____ + energy

18 Heart

Date for completion: / /

Parent sig: _____
Teacher sig: _____

SP1 Unit 2.12

1 Label parts **a** to **k** in this drawing.

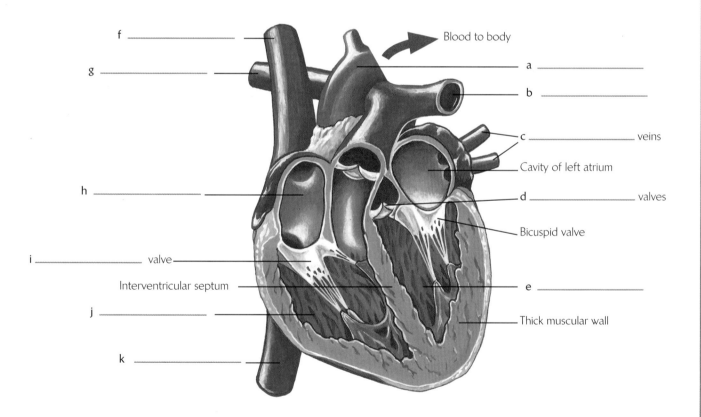

f _____
g _____
Blood to body
a _____
b _____
c _____ veins
Cavity of left atrium
h _____
d _____ valves
Bicuspid valve
i _____ valve
Interventricular septum
e _____
j _____
Thick muscular wall
k _____

2 Add arrows to the above drawing, to show the direction of blood flow.

ISBN: 9780170214650

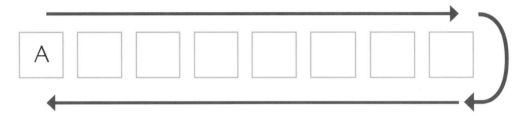
3 Eight events are listed below in the wrong sequence. Arrange them into the correct sequence, starting with A.

A blood enters the left atrium

B blood high in oxygen travels along the pulmonary vein

C left atrium contracts and forces blood into the left ventricle

D the right ventricle contracts, forcing blood along the pulmonary artery

E blood returns from the body, carrying very little oxygen

F the left ventricle contracts and forces blood into the aorta

G blood travels to the lungs where it picks up oxygen and unloads CO_2

H blood trickles through the capillaries, unloads oxygen, then is collected together in the veins before its return to the heart.

A							

19 Heart transplants

Date for completion: / /   Parent sig: _____ Teacher sig: _____

SP1 Unit 2.12

Heart transplants are performed on patients with end-stage heart failure. The most common procedure is to take a working heart from a recently deceased organ donor and implant it into the patient. Post-operation survival periods now average 15 years.

A typical heart transplantation begins with a suitable heart being located from a brain-dead donor. The transplant patient is contacted and instructed to attend the hospital for pre-surgical medication. At the same time, the heart is removed from the donor and inspected by a team of surgeons to see if it is in a suitable condition to be transplanted. Occasionally it will be deemed unsuitable.

This can often be a very distressing experience for a patient, and they will usually require emotional support before being sent home. The patient must also undergo many emotional, psychological and physical tests to make sure that they are in good mental health and will make good use of their new heart. The patient is also given medication so that their immune system will not reject the new heart.

If the donor heart passes inspection, the patient is taken into the operating room and given a general anaesthetic. The procedure

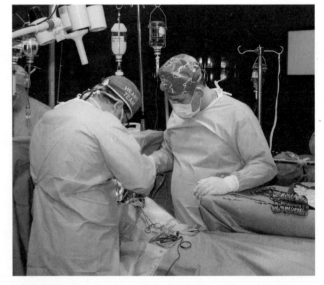

begins with the surgeons opening the rib cage. The patient is then attached to a bypass heart-lung machine. The failing heart is removed, the donor heart is fitted into place, and the main blood vessels are sewn in place. The new heart is restarted, the patient is removed from the bypass machine, and the chest cavity is closed.

Post-operative complications include infection as well as the side-effects of the immuno-supressive medication. Since the transplanted heart originates from another organism, the recipient's immune system may attempt to reject it.

1 Underline key facts in the heart transplant article on the previous page.

2 Write a short summary on the subject in 10 to 12 bullet points. Avoid using words you not sure about.

_____

_____

_____

_____

_____

_____

_____

_____

_____

_____

_____

_____

20 Joints and muscles

Date for completion: / /

Parent sig: _____
Teacher sig: _____

SP1 Unit 2.13

The following sentences have been scrambled, and the three main parts do not match. Rewrite all the sentences in the spaces below, so that each makes a correct statement of fact.

	Subject	Verb	Object
1	Synovial fluids	bends	the whole arm.
2	Ligaments	lifts	joints from dislocating.
3	Tendons	lubricate	the elbow.
4	The bicep muscle	straightens	muscle to bone.
5	The tricep muscle	attach	the spaces inside joints.
6	The deltoid muscle	prevent	the elbow.

Correct statements:

1 Synovial fluids _____

2 Ligaments _____

3 Tendons _____

4 The bicep muscle _____

5 The tricep muscle _____

6 The deltoid muscle _____

ISBN: 9780170214650

Date for completion: / /

Parent sig: _____
Teacher sig: _____

SP1 Unit 2.13

Life Science

1 Give the correct names of bones ABCD.

A _____

B _____

C _____

D _____

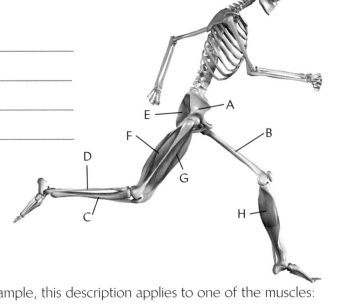

2 Describe the action of muscles EFGH. For example, this description applies to one of the muscles: *Pulls the femur and the whole leg backwards.* Decide which muscle this action fits, and write similar descriptions for the others.

E _____

F _____

G _____

H _____

I Using a different colour, carefully draw muscle I, which pulls the whole arm back.

3 Give the correct names of structures A to G in the diagram below.

Which muscle bends the elbow? Letter: _____ Name: _____

4 Using a different colour, carefully draw in the deltoid muscle (H) which lifts the whole arm up and sideways. Make it clear to which bones it is attached at each end.

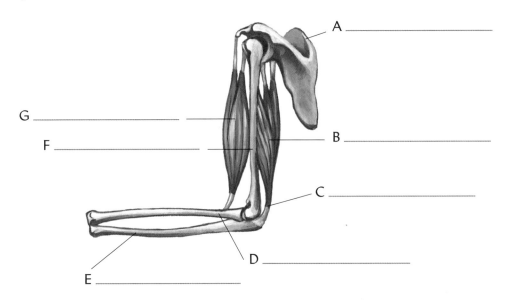

A _____

G _____

F _____

B _____

C _____

D _____

E _____

ISBN: 9780170214650

SP1 Unit 2.14

Life Science

This diagram represents a cross-section through a mammal eye, but is incomplete.

1   Label parts A to G.

2   Carefully draw in the iris, pupil, lens, and the ligaments that hold the lens in place. Label the four parts you have added.

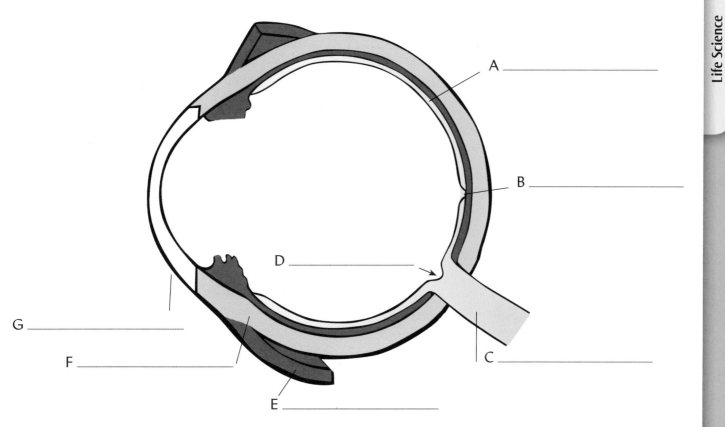

A _____

B _____

D _____

G _____

F _____

E _____

C _____

SP1 Unit 2.14

1   Each eye has a blind spot that cannot see images. This blind spot is the area surrounding the optic nerve. This activity helps you to 'map' your blind spot.

Step A: Cover your left eye and using your right eye only, look steadily at the X on the following page, from a fixed distance of about 25 cm.

Step B: Move a sharp-pointed pencil steadily from X towards the right side of the page. Somewhere near the circle, the tip of your pencil will disappear, then reappear when you move it further to the right. Try not to move your gaze from X. The points of disappearing and reappearing mark the boundary of your right eye's blind spot.

Step C: Continue moving your pencil up and down, and left and right, in the general vicinity of the circle. In every place where the pencil tip becomes invisible, make a mark on the paper. Eventually these marks will show the area that your right eye is 'blind' to.

ISBN: 9780170214650

X

2 Explain why you are not normally aware of your blind spot.

_____

_____

_____

_____

_____

_____

_____

CHALLENGE

3 From your own results, suggest at least one more blind spot question that you could easily check out by testing your own eyes.

_____

_____

_____

_____

_____

_____

_____

_____

_____

_____

_____

ISBN: 9780170214650

1 Complete the sentences below by writing in words chosen from this list: *brain, axon, reflex, spinal, axon, hormones, synapse, 105 m/s, neurones.*

Different cells and regions in your body are constantly communicating with others in different ways

mainly through _____ (1), and through nerve cells, also known as

_____ (2). Many kinds of these cells have a long thin extension

called an _____ (3). Electrical signals travel along these axons at

high speed, about _____ (4). There are brief delays in each gap

or _____ (5) between cells. The simplest kind of nerve response

is a _____ (6) , such as the protective blink of your eye. All

reflexes involve nerve signals being sent via the _____ (7) or the

_____ (8) cord.

2 This diagram gives a simple picture of how one reflex works: your iris closing down in response to bright light. Plan a simple test on this iris reflex based on the following aim.

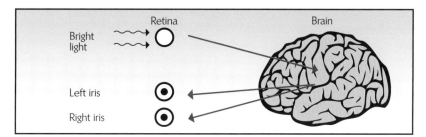

Aim: To find out if your left and right iris are independent, or if they are linked and both react to light together.

Your method: Describe your plan.

_____

_____

_____

_____

_____

_____

_____

Your results: After checking, describe what you found.

_____

_____

_____

ISBN: 9780170214650

This drawing describes one way of measuring eye-hand reaction time. It is based on how fast the ruler falls, and how quickly you can grab it when your partner lets go. Alex and Ben are competitive, and want to see who has the fastest reactions.

Here are their results for seven tests each. Each result was measured to the nearest centimetre (cm).

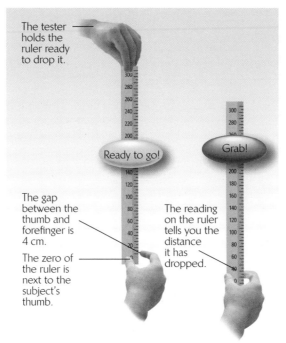

The tester holds the ruler ready to drop it.

Ready to go!

Grab!

The gap between the thumb and forefinger is 4 cm.

The zero of the ruler is next to the subject's thumb.

The reading on the ruler tells you the distance it has dropped.

Results:

Trial number →	1	2	3	4	5	6	7
Alex	11 cm	12 cm	11 cm	12 cm	12 cm	29 cm	13 cm
Ben	13 cm	11 cm	0 cm	12 cm	14 cm	11 cm	14 cm

1 Calculate the average distance the ruler fell before it was caught: by Alex _____ cm, and by Ben _____ cm.

2 Two of the results are probably not fair or reliable. Suggest which two?

_____

3 Suggest what caused these results to happen.

_____

_____

4 If you exclude these two unfair results, calculate the new averages. Alex _____ cm, and Ben _____ cm.

Life Science

5 Draw a graph based on the following figures. They show how far an object will fall in each 10 millisecond interval. Remember to give your graph a title, and to label the axes.

Time in milliseconds (ms)	130	140	150	160	170	180	190	200
Average distance of fall (cm)	8.3	9.6	11.1	12.5	14.2	15.9	17.7	19.6

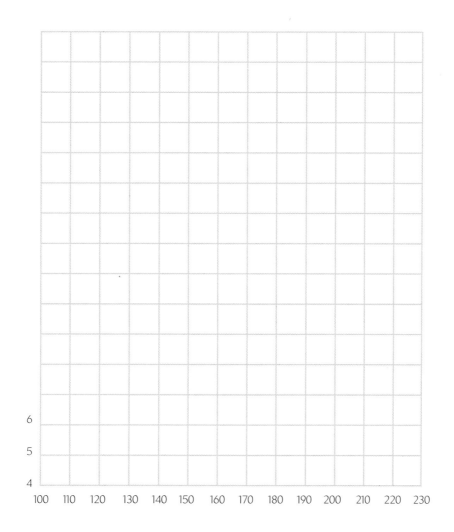

6 Use your graph to predict the average eye-hand reaction times. For Alex _____

For Ben: _____ Who was quickest? _____

7 Use the graph to find how far an object will fall in:

0.1 s _____

0.2 s _____

145 ms _____

ISBN: 9780170214650

26 Plant structure

Date for completion: / /

Parent sig: _____

Teacher sig: _____

SP1 Unit 2.16

Complete the diagram below by placing each of the following labels in its correct place:
primary root, root hairs, stem, flower, petiole, leaf blade, terminal bud, root system, shoot system.

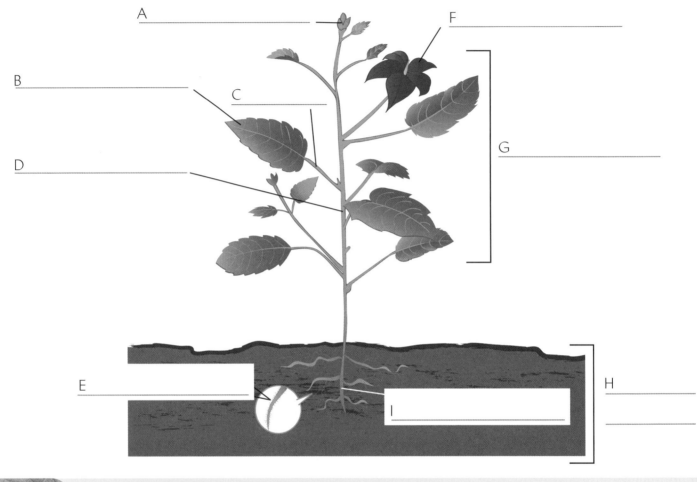

A _____

F _____

B _____

C _____

D _____

G _____

E _____

H _____

I _____

27 Growth experiment

Date for completion: / /

Parent sig: _____

Teacher sig: _____

SP1 Unit 2.16

For her plant project, Katie planted three containers with sunflower seeds.

Container A: 200 g ordinary soil with no added fertiliser.

Container B: 200 g ordinary soil with ¼ teaspoon fertiliser mixed in at the start.

Container C: 200 g ordinary soil with 3 teaspoons fertiliser mixed in at the start.

She photographed the plants every week and recorded her results as follows:

	Start	Week 1	Week 2	Week 3
A	12 seeds	10 just starting	12 growing well	12 quite tall
B	12 seeds	10 just starting	12 growing well	12 tall, maybe taller than A
C	12 seeds	5 just starting	2 OK, 3 look sick	Only 1 alive, 4 definitely dead

Aim: What was the aim of Katie's experiment?

_____

_____

_____

Method: According to her notes, list three variables Katie kept the same for containers A, B and C.

_____

_____

_____

_____

Suggest at least three other variables she probably tried to keep the same for A, B and C, but are not mentioned in the description above.

_____

_____

_____

_____

_____

Which trial (A, B or C) was the 'control', the one she did for the sake of comparison?

_____

_____

Conclusion: Suggest a conclusion, based on her results.

_____

_____

_____

Evaluation: Suggest what Katie could have done to improve the way she recorded her results.

_____

_____

_____

_____

_____

CHALLENGE

ISBN: 9780170214650

Date for completion: / /
Parent sig: _____
Teacher sig: _____

Life Science

1 Complete the crossword puzzle.

ACROSS

1 Contains the female reproductive cells that can become seeds .
3 The parts of a flower that attracts insects by their colour
6 This connects the stigma and the ovary. The pollen tube grows down this.
7 The combined female part of the flower .
8 The stalk of the stamen. This holds the anther.

DOWN

2 The part of the stamen that produces pollen.
3 Grows down towards the egg.
4 The combined male parts of the flower.
5 The top of a flower's pistil. This is where the pollen arrives.

2 Label the drawing, parts **A** to **I**.

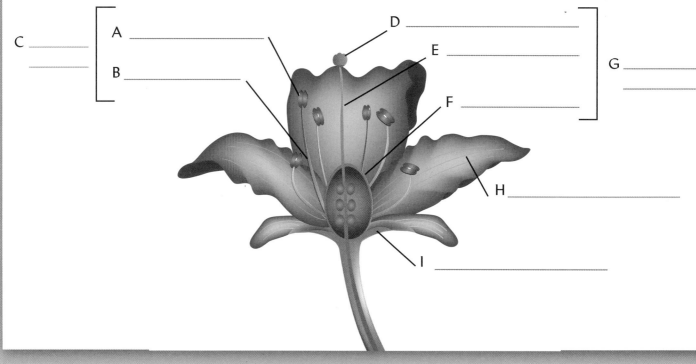

C _____

A _____

B _____

D _____

E _____

F _____

G _____

H _____

I _____

ISBN: 9780170214650

Plant reproduction

Date for completion: / /

Parent sig: _____
Teacher sig: _____

1 Use words from this list to label features A to G: *runner, germination, asexual reproduction, cutting, pollination, sexual reproduction, seedling.*

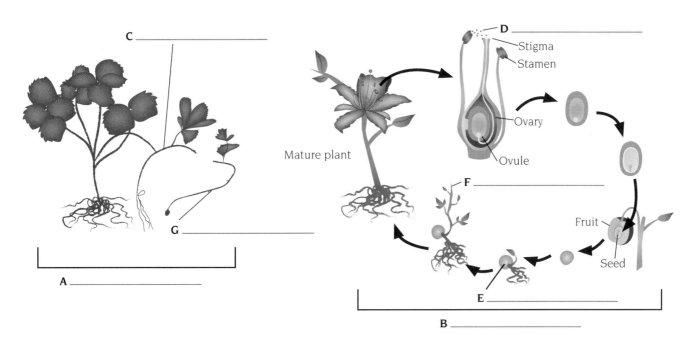

C _____

G _____

A _____

Mature plant

D _____

Stigma

Stamen

Ovary

Ovule

F _____

Fruit

Seed

E _____

B _____

2 Use the diagram to describe step-by-step the life cycle of a plant that uses sexual reproduction.

First, a seed germinates. _____

Next, ... _____

_____

_____

_____

_____

_____

_____

_____

_____

_____

_____

_____

ISBN: 9780170214650

Life Science

Insects and wind are the two main methods that flowers use to spread pollen to other flowers of the same kind. For each method:

- name one example of an actual plant that uses this method.
- describe special features of the flowers, and/or the pollen grains.
- make a drawing to help your description.

Method 1: Insects

_____
_____
_____
_____
_____
_____
_____
_____
_____

Method 2: Wind

_____
_____
_____
_____
_____
_____
_____
_____
_____
_____

31 Seed spread

Date for completion: / /
Parent sig: _____
Teacher sig: _____

SP1 Unit 2.18

Plants have many different methods of getting seeds away from their parents. These methods include using **wind**, **water**, **birds**, and **mammals**. The word for spreading and travelling is **dispersal**. If a seed is lucky enough to be dispersed to a suitable place, it can start to germinate and grow.

For each of the seeds shown below, suggest:

A Which kind of dispersal method the plant is using.

B Ways in which the seeds are adapted to that method of dispersal.

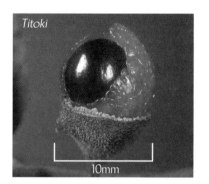

Titoki

10mm

_____
_____
_____
_____
_____

Thistle

_____
_____
_____
_____
_____

Coconut

_____
_____
_____
_____
_____

Bidibid

_____
_____
_____
_____
_____

Life Science

ISBN: 9780170214650

Date for completion: / /
Parent sig: _____
Teacher sig: _____

SP1 Unit 2.18

The following events are all stages in the life cycle of a plant that reproduces using sex.

Arrange these events (A to J) in a life cycle, in the order they occur. Your cycle can begin at any point, because a cycle does not have a beginning or an end.

A seed develops inside the ovary of the flower
B pollen arrives on the stigma
C seed is taken from the parent plant, or falls off
D seeds are spread by birds, insects, animals, wind or water
E the eggs are fertilised
F seed lands in a suitable place, then germinates
G plant matures
H new plant begins to grow
I flowers develop
J pollen tubes grow towards the eggs

Life Science

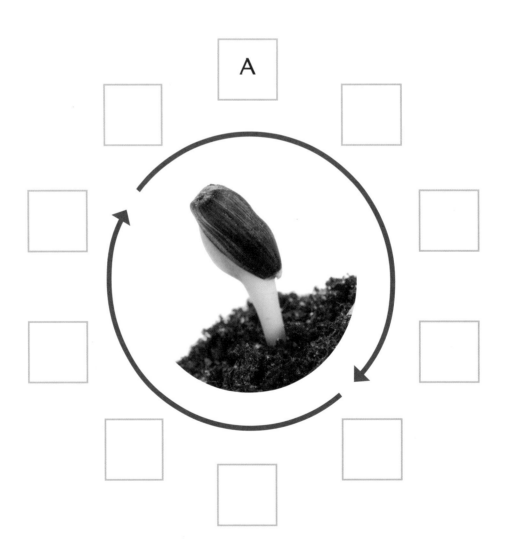

A

SP1 Unit 2.19

Most plants can reproduce by two methods that pass on genes to the next generation:

- sexual methods
- asexual methods (non-sexual).

Using the information from previous activities, compare and contrast these two types of reproduction in this T-diagram. Use the following points:

A   Number of parents needed?   B   Offspring identical to parent?
C   Which type of reproduction is faster?   D   Advantages compared to the other method?
E   Disadvantages compared to the other method?   F   Same species, or two species?

Sexual methods	Non-sexual methods

34 **Photosynthesis experiments**

SP1 Unit 2.20

1   How can we check whether or not photosynthesis is happening in a plant? Photosynthesis makes sugar and starch. When starch meets iodine it always turns black, and this simple chemical test can tell us when photosynthesis has been happening. But to make a black stain visible you first have to burst the cells by boiling, then dissolve out chlorophyll – the green colour. To test leaf starch, we usually carry out practical steps ABCD that follow. Complete the following statements that summarise the process.

ISBN: 9780170214650

Life Science

Step A Leave the plant in bright sunlight for at least three hours. Reason:

_____

_____

Step B Remove a leaf and place it in boiling water for two minutes. Reason:

_____

_____

Step C Place the leaf in clear alcohol solution and boil it for two minutes. Reason:

_____

_____

Step D Gently wash the leaf, then pour some iodine solution over it. Reason:

_____

_____

2 Leaf X was partly covered with cardboard for 12 hours in bright sunlight. Use a pencil to shade in the regions that you predict will go black after a starch test.

3 Leaf Y is **variegated**, which means it is naturally partly green and partly not. Use a pencil to shade in the regions that you predict will go black after a starch test.

4 Label the arrows on this drawing to show the inputs and outputs of photosynthesis.

A_____

B_____

Photosynthesis

C_____

D_____

E_____

ISBN: 9780170214650

Date for completion: / /

Parent sig: _____

Teacher sig: _____

SP1 Unit 2.20

1 Below is a diagram of a leaf cross section. Label the diagram using terms from this list, and also use them to answer the questions below: *palisade cells, epidermis, guard cell, air spaces, stoma, vein, waxy cuticle.*

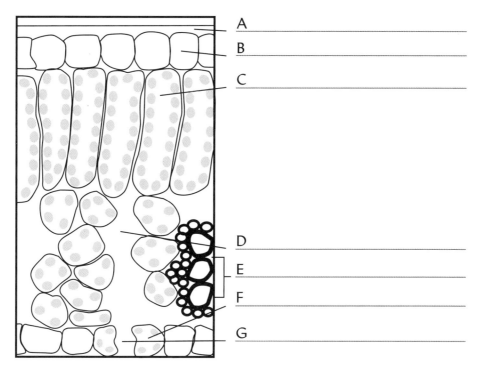

A _____

B _____

C _____

D _____

E _____

F _____

G _____

2 Colour the above diagram. Make the chloroplast 'dots' = green; cuticle = yellow; vein cells = blue; other cells = pale grey pencil.

3 Where in the leaf does all photosynthesis take place?

_____

4 What is the function of guard cells?

_____

_____

5 Name one substance that enters a leaf through its stomata.

_____

6 Name two substances that exit the leaf through the stomata.

_____

7 Epidermis cells are transparent. Suggest the function of this.

_____

_____

8 Leaves have a waxy cuticle. Suggest the function of this.

_____

_____

9 Name one substance that travels into the leaf through the veins.

_____

10 Name two substances that travel out through the veins of a leaf.

_____

Life Science

ISBN: 9780170214650

Life Science

Photosynthesis is a process that green plants use to make their own food.

An experiment was conducted with two marigold plants. One was put outside on a windowsill and the other was placed inside a cupboard. Both plants were watered every day. After five days, the plant in the cupboard was taken out and placed next to the plant that had been on the windowsill. The plant from the cupboard was yellowy and some of the leaves had fallen off, while the plant from the windowsill looked green and healthy.

1 What does this experiment demonstrate to you about the requirements of plants?

_____

_____

2 What else do plants require in order to survive? Name three requirements.

_____

_____

_____

3 Use your answer from the questions above to write out a word equation for photosynthesis in the boxes below.

[box]

[box] + [box] → [box] + [box]

4 In 7 to 15 words, explain why photosynthesis is so important to plants.

_____

_____

_____

_____

_____

Date for completion: / /

Parent sig: _____

Teacher sig: _____

SPT Unit 2.20/2.22

Life Science

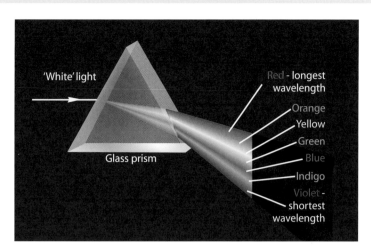

Title: Experiment to test which light colours are most useful to a plant for photosynthesis.

Hypothesis: (Based on what colour or colours you think the plant will need.)

_____

_____

What factor (independent variable) will you change in your experiment? Plan at least three trials.

_____

_____

_____

List at least three factors you will keep the same in your three trials, so that it is a fair experiment:

_____

_____

_____

_____

Write a step-by-step method of how you would carry out this experiment. Do drawings if they help.

_____

_____

_____

_____

_____

_____

_____

_____

_____

_____

ISBN: 9780170214650

Life Science

Waikeri was planning a project experiment on osmosis. She cut two pieces of carrot to the same size. She put one of the pieces (A) in a glass of salty water, and the other piece (B) in a glass of ordinary tap water. A few hours later, the salt-water carrot looked smaller and felt rubbery. The carrot in the fresh water was firm and crunchy.

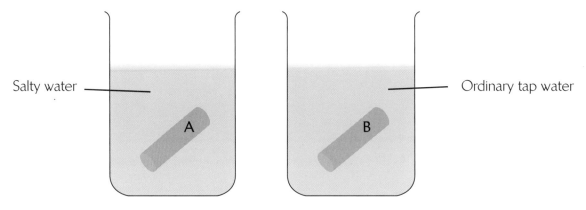

Salty water — A

Ordinary tap water — B

1 Using your knowledge of osmosis, explain what happened inside carrot pieces A and B.

_____
_____
_____
_____
_____
_____
_____
_____
_____

2 Suggest at least one step she could have taken to improve her results.

_____
_____

CHALLENGE

3 Draw arrows showing which way the water moves between water and carrot. The pictures represent the different proportions of salt and water particles. (There are of course millions of particles. The broken line represents cell membranes which let water molecules pass through.)

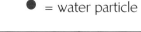

● = water particle   ● = salt particle

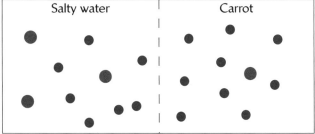

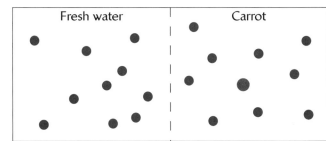

Date for completion: / /

Parent sig: _____

Teacher sig: _____

SP1 Unit 2.16–2.22

The following sentences have been scrambled. Rearrange them to write seven true statements about plants.

Seeds	have a large surface area	and eventually make seeds
Leaves	contain sex organs	to absorb water
Flowers	are long and narrow	to absorb sunlight, and make sugars
Buds	have a large surface area	to spread plants to new areas
Roots	carry sugar solutions	and carry water from roots to leaves
Phloem cells	are growth points	from leaves to growing parts
Xylem cells	are made in big numbers	mainly on the stem

Correct statements:

Seeds

_____

_____

Leaves

_____

_____

Flowers

_____

_____

Buds

_____

_____

Roots

_____

_____

Phloem cells

_____

_____

Xylem cells

_____

_____

Life Science

ISBN: 9780170214650

Date for completion: / /

Parent sig: _____
Teacher sig: _____

SP1 Unit 2.16–2.22

Life Science

1 Some tree *structures* resemble an upside-down version of human lungs.

The tree trunk looks like

_____

The branches look like

_____

The leaves look like

_____

In what way is the *function* of leaves similar to the *function* of lung alveoli? Explain.

_____

_____

_____

2 Complete this table.

Structure	Function
	Reproducing part of the plant
Roots	
	Gives the plant support above the ground
	The part of the plant responsible for photosynthesis
Root hairs	

ISBN: 9780170214650

Chemical Science

1 States and particles

Date for completion: / / Parent sig: _____ Teacher sig: _____

1 Complete the labelling of the diagram below, by writing the following words in their correct places: *solid, liquid, condensing, melting, evaporating, vaporising, solidifying, freezing, boiling.*

   In some cases two words could be placed next to one arrow.

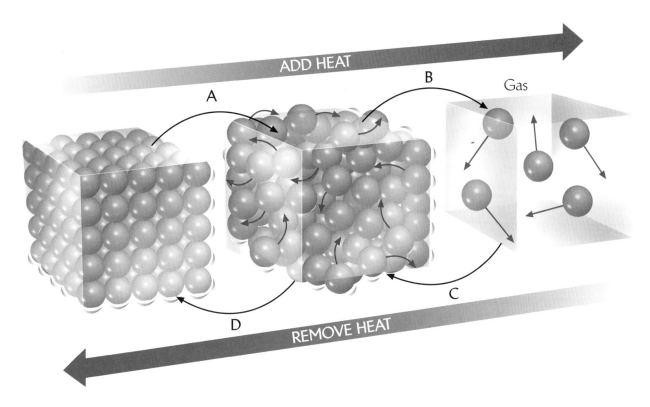

2 Add a two-headed arrow labelled **sublimating** to the drawing above.

3 Explain the difference between 'melting' and 'sublimating'. Name two substances that can sublimate.

_____

_____

_____

_____

_____

CHALLENGE

ISBN: 9780170214650

Date for completion: / /

Parent sig: _____

Teacher sig: _____

SP1 Unit 3.1

Complete the table below.

	Solid state (or phase)	Liquid state (or phase)	Gas state (or phase)
Name any one example of a substance in this state.			
Can its shape be easily changed?			
Does it have a **definite volume** that changes very little?			
Does it have a **definite shape**?			
Compare the spacing and arrangement of its particles.	Close-packed, regular arrangement		
Compare the average particle **energy** and speed.	Slow, low energy	Faster, higher energy	

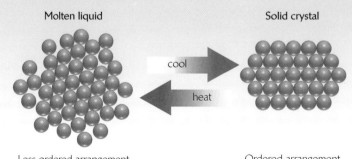

Molten liquid   cool →   Solid crystal

     ← heat

Less ordered arrangement   Ordered arrangement

Chemical Science

ISBN: 9780170214650

3 Melting and boiling

Date for completion: / / Parent sig: _____ Teacher sig: _____

SP1 Unit 3.1

1 Most pure substances melt and boil and freeze at exact temperatures – as long as the pressure conditions have not changed. Give the temperatures for each of the changes below, remembering to write the correct units.

A freezing point of pure water _____

B melting point of pure ice _____

C boiling point of water _____

D temperature at which steam condenses to water _____

E melting point of gold _____

F temperature at which liquid gold solidifies _____

G temperature at which gold vaporises _____

H temperature at which CO_2 gas sublimates to a solid _____

2 The boxes below represent the same amount of a substance (aluminium) in three different states: solid, liquid, gas. Draw particles (atoms in this case) in the liquid and gas boxes. Note: the particle size does not change.

Solid Al   Liquid Al   Al gas

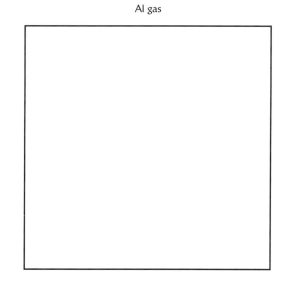

4 How do we know?

Date for completion: / / Parent sig: _____ Teacher sig: _____

SP1 Unit 3.1/3.14

Particle theory describes what happens when solids melt and gases condense. But how can we be sure? Atoms and molecules are too small to see, so we have to rely on indirect evidence. In the table on the following page, match up theory statements and everyday evidence by writing the letter for each item of evidence A to G in the empty box matching a particular theory statement.

Chemical Science

ISBN: 9780170214650

Theory statement	
In a gas, the particles are far apart.	
Particles in a liquid are almost as close together as in solids.	
Solid particles are in regular patterns, but liquid particles not.	
The particle energy in gases is much greater than in liquids.	
The particle energy in liquids is much greater than in solids.	
Particles in a solid are as close together as they can be.	
Particles move faster when the temperature is higher.	

Everyday evidence	
A	A lump of ice has a fixed shape, but when it melts the water has no fixed shape.
B	Liquid rises up the tube of a thermometer when its temperature rises.
C	It takes a lot of heat to melt solid lead.
D	You can hammer a piece of lead into a new shape, but you can't make it smaller.
E	You can easily squash an inflated balloon, and it will always rebound.
F	It takes enormous pressure to squash 1 L water into a slightly smaller volume.
G	Steam at 100 °C can harm you much more seriously than water at 100 °C.

Evidence: facts or signs showing that something exists or is true.
Theory (one meaning): a big idea that explains many facts.

5 Particle theory

Date for completion: / /   Parent sig: _____
Teacher sig: _____

SP1 Unit 3.1/3.14

Explain facts A to G by using particle theory.

A A plastic bottle full of **air** and with the top screwed on tight gets squashed smaller when you take it to the bottom of a swimming pool. Reason:

_____

_____

_____

B A plastic bottle full of **water** and with the top screwed on tight does not get squashed when you take it to the bottom of a swimming pool. Reason:

_____

_____

_____

C When liquid candlewax becomes solid, it shrinks slightly. Reason:

_____

_____

_____

ISBN: 9780170214650

D If you pour any liquid into a cup, it will fill the cup and take the shape of the cup. Reason:

_____

_____

_____

E If you try to cook an unbroken egg in a microwave oven, the egg will explode. Reason:

_____

_____

_____

F When it starts snowing the air temperature warms slightly; but when snow is melting the air becomes much colder. Reason:

_____

_____

_____

G When you put a pot of water on the stove it may take only a few minutes for the water to reach 100 °C, but it takes much longer for the water to boil away completely. Reason:

_____

_____

_____

_____

6 Hidden heat

Date for completion: / /   Parent sig: _____

Teacher sig: _____

SP1 Unit 3.3

Olivia and Hemi's aim is to heat lumps of ice over a Bunsen burner, then measure the temperature of the ice and melted water every minute. Their ice was taken from a very cold deep-freeze. Their setup is shown in the drawing, and their results are in this table.

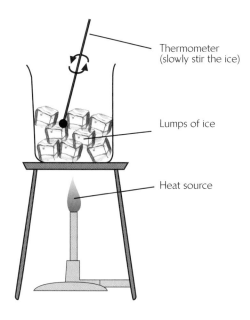

Thermometer (slowly stir the ice)

Lumps of ice

Heat source

Time (minutes)	Temp. (°C)	What we saw
start	-9	all ice
1	-2	all ice
2	0	mostly ice, some liquid water
3	1	all water
4	21	water
5	14	water
6	59	water
7	-	water
8	86	water with some bubbles
9	95	more bubbles starting
10	99	lots of bubbles
11	100	bubbling fast
12	100	bubbling fast
13	100	bubbling fast
14	100	bubbling fast

ISBN: 9780170214650

1 Draw a line graph to represent these figures. Results (temperature in this case) traditionally go on the vertical axis. Use the TADPL guidelines on page 15.

0

0   1   2   3   4   5   6   7   8   9   10   11   12   13   14

Time (minutes)

2 One of their temperature measurements was recorded wrongly. Which one? _____

Suggest what the correct measurement probably was. _____

3 They both missed the temperature in the seventh minute. Use the graph to estimate the probable correct

figure. _____

4 At which two temperatures was there a change of state? _____

5 What substance was inside the bubbles? _____

ISBN: 9780170214650

Chemical Science

6 After 11 minutes the temperature did not rise any more, even though the flame was supplying heat at a steady rate. Explain where this flame heat went.

_____

_____

7 'Latent' means 'hidden'. Explain why the heat needed to melt the ice is described as 'hidden'.

_____

_____

_____

8 Olivia and Hemi continued their experiment for another 10 minutes. Predict what they probably observed:

a about temperature.

_____

b about water level (give a reason for predicting this).

_____

_____

_____

7 Number puzzle

Date for completion: / /   Parent sig: _____   Teacher sig: _____

SP1 Unit 3.1/3.3

Inside this number puzzle grid are nine commonly-used words about changing states. First, use the clues below to find what these words are. (The first has been done for you.) Next, find and circle the other eight words in the number puzzle. Words can be read horizontally or vertically.

17	1	16	19	15	9	8	19	19	18	12
25	23	21	20	5	24	18	15	15	1	5
11	5	17	5	9	7	22	22	12	22	22
24	12	7	1	10	1	24	22	9	13	1
3	25	14	13	17	19	4	21	4	18	16
20	23	23	10	26	24	14	25	19	5	15
13	1	6	18	5	5	26	5	10	9	18
5	12	14	11	16	9	23	3	17	8	1
12	9	10	1	2	23	8	3	18	2	20
20	11	9	1	17	10	5	21	17	6	5
9	4	24	17	1	18	1	11	21	15	7
24	16	12	8	1	4	20	10	22	10	1
23	8	9	12	9	17	21	9	4	3	26
1	2	11	6	23	14	24	10	16	11	9
3	15	14	4	5	14	19	5	24	16	13

Clues

1-A, 2-B, 3-C, 4-D, 5-E, 6-F, 7-G, 8-H, 9-I, 10-J, 11-K, 12-L, 13-M, 14-N, 15-O, 16-P, 17-Q, 18-R, 19-S, 20-T, 21-U, 22-V, 23-W, 24-X, 25-Y, 26-Z

8 5 1 20 _____ H E A T _____

19 15 12 9 4 _____

19 20 5 1 13 _____

12 9 17 21 9 4 _____

3 15 14 4 5 14 19 5 _____

6 18 5 5 26 5 _____

5 22 1 16 15 18 1 20 5 _____

13 5 12 20 _____

7 1 19 _____

ISBN: 9780170214650

8 Exothermic, endothermic

Date for completion: / /

Parent sig: _____
Teacher sig: _____

SP1 Unit 3.1/3.3

For each of the following 11 situations:
- give the change of state (if none, say so)
- identify whether the change is **endothermic** (takes in heat energy), or **exothermic** (gives out heat).

Add one more 'physical change' situation of your own choice to complete line 12.

	Situation	Change of state?	Exothermic, endothermic or neither?
1	Ice melting	Solid to liquid	
2	Water boiling		
3	Snow forming		
4	Sweat evaporating		
5	Sugar dissolving		
6	Clouds forming		
7	Wind blowing		
8	Stone falling		
9	Door closing		
10	Ice freezing		
11	Snow melting		
12			

ISBN: 9780170214650

9 Solutions and crystals

Date for completion: / /

Parent sig: _____

Teacher sig: _____

SP1 Unit 3.4/3.5

Write the missing word in each of the numbered spaces. Choose from these words:
suspension, soluble, concentrated, water, petrol, particles, crystals, grease, insoluble, super-saturated, dilute, slowly, saturated, colloid, sugar.

When sugar is dissolved in water, the _____ (1) is the solvent, and the

_____ (2) is the solute. When you use petrol to remove grease from engine parts,

the solvent is the _____ (3) and solute is the _____ (4).

Some chemicals are very _____ (5), like sugar in the water. Other

substances won't dissolve; they are completely _____ (6), like sand in water.

Clay particles are too big to dissolve in water. When you stir some clay into water you get a

cloudy-looking _____ (7) which can't go through filter paper, or perhaps a

_____ (8) which can go through filter paper.

It's easy to change the amount of solute. If you stir a teaspoon of salt into a litre of water, the solution

is very _____ (9). But 200 grams of salt in a litre of water would make a more

_____ (10) solution. If you add more and more salt, eventually no more will dissolve:

the solution has become _____ (11). If you take this kind of solution and let some

of the water evaporate, it will become _____ (12), and eventually will start to grow

_____ (13).

Crystals have a regular shape because

the _____ (14) are

arranged in regular patterns. Crystals

tend to become bigger if they grow more

_____ (15).

Solvent evaporates.

Liquid solvent and solute particles are mixed.

Solute particles arrange themselves in crystals.

Chemical Science

ISBN: 9780170214650

10 Elements crossword

Date for completion: / /
Parent sig: _____
Teacher sig: _____

SP1 Unit 3.6

ACROSS

1 Element number 15
5 Element number four
6 Trains run on it
7 Helps protect steel from rusting
9 You can't go more than three minutes without it
10 Used in mag alloy wheels
11 In the air and in proteins
13 Makes up almost 1% of the air
15 A light metal
19 Needed for your thyroid gland
21 Plenty of this in your bones
25 Its name was originally 'pot-ash'
27 Used to sterilise swimming pools
28 Rocks are mostly this element
29 A metal used in car batteries

DOWN

2 The smallest atom of all
3 Sometimes used to fill balloons
4 Best stay away from its nuclear reactions
8 Electricity runs through it
12 Used in fluorescent lighting
14 Used in jewellery throughout the ages
16 Light metal, used in some batteries
17 Liquid metal
18 Used in shiny coins
20 A yellow powder
22 Sounds like a boring element
23 Plenty of this in wood
24 A tiny bit of it helps strengthen your teeth
26 Shiny and valuable

Date for completion: / /

Parent sig: _____
Teacher sig: _____

SP1 Unit 3.6

1 Count the total number of metal elements shown in blue in the periodic table below _____

2 How many element symbols start with the letter C? _____. Write the name and symbol of each. _____

_____

_____

3 Write the names and symbols of two elements starting with letter I.

_____

4 Write the names and symbols of two elements starting with letter A.

_____

5 Name the element that makes up 78% of the air. _____

6 Name the element that makes up about 20% of the air. _____

7 The elements arranged on the far side of the table (group 18) are usually described as 'inert' gases because they hardly ever react with other elements. Give the names and symbols of the first three

_____

What do you notice about their atomic numbers?

_____

8 The metals arranged on the left side of the table (group 1) are usually described as alkali metals, because they react with water to give alkali solutions. Give the names and symbols of the first three.

_____

What do you notice about their atomic numbers?

_____

9 Human bodies, as with most other living things, consist more than 99% of the elements C, H, O, P, K, N, S, P, Ca, Mg, Na, Cl and Fe. Put a circle around each of these element symbols in the periodic table. Generally, what do you notice about the positions on the table of these elements?

_____

_____

Chemical Science

CHALLENGE

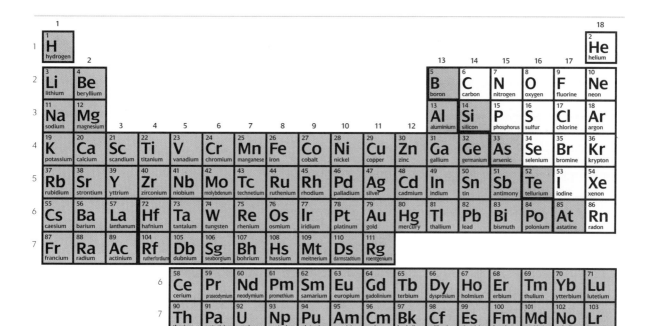

ISBN: 9780170214650

12 Alphabet elements

Date for completion: / /
Parent sig: _____
Teacher sig: _____

SP1 Unit 3.6

1 This is another way of helping you memorise the names and symbols of the first 20 chemical elements, plus 10 more. Use the periodic table to complete the tables below.

Symbol	Name	Number
Al	aluminium	13
Ag		
Ar		
Au		
B		
Be		
C		
Ca		
Cl		
Cu		
F		
Fe		
H		
He		
Hg		

Symbol	Name	Number
I		
K		
Li		
Mg		
N		
Na		
Ne		
Ni		
O		
P		
Pb		
S		
Si		
U		
Zn		

2 Many chemical symbols are based on English-language names and are easy to remember, like S for sulfur. However some symbols come from other languages – and it can be useful to know this when memorising symbols. Complete the table below.

Original name	Number	Symbol	English name
natrium (in Latin)	11		
fer (the French word for iron)	26		
kalium (in Latin)	19		
argentum (the Latin word for silver)	47		
aurum (the Latin word for gold)	79		
hydro-argentum (means water-silver)	80		
plumbum (the old word for lead)	82		
wolfram (a German name)	74		

13 Elements, compounds, molecules

Date for completion: / /

Parent sig: _____
Teacher sig: _____

Three different types of atoms are shown below, colour-coded as red, blue and green. Classify the diagrams of substances A to F by writing the following words in the correct places in the table: *mixture, element, compound, pure substance, molecules, single atoms.*

Some descriptions may need two or more words.

	Type of substance (mixture, element, compound, pure substance)	Type of particles (molecules, single atoms, or both)
A		
B		
C		
D		
E		
F		

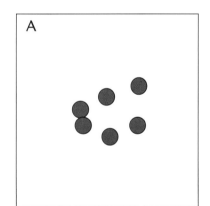

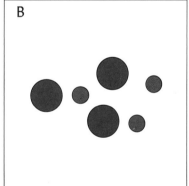

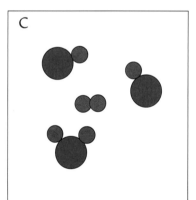

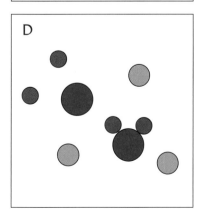

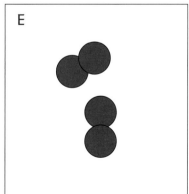

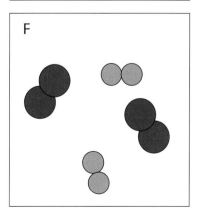

Chemical Science

ISBN: 9780170214650

Date for completion: / /

Parent sig: _____

Teacher sig: _____

SP1 Unit 3.7

Chemical Science

1 Classify each of these 14 substances as a single element, a single chemical compound, or a mixture of compounds and perhaps elements as well: *water, seawater, air, oxygen, glucose, sugar, milk, tomato sauce, petrol, aluminium, diamond, copper sulfate, rust, concrete, ordinary salt.*

Write each name in its correct column.

Single elements	Single chemical compounds	A mixture of different compounds, and perhaps elements as well
Not sure:		

2 List three ways in which mixtures are different to compounds.

_____

_____

_____

_____

_____

_____

_____

Copper (Cu)

Sulfur (S)

Copper sulfate crystals (CuSO$_4$)

ISBN: 9780170214650

Date for completion: / /
Parent sig: _____
Teacher sig: _____

SP1 Unit 3.7

The name of a compound can tell you what elements make it up. In many cases, two element names are simply joined together and '**-ide**' put at the end. If the name ends in '**-ate** ' or '**-ite**', this tells you the compound contains oxygen atoms as well as the other elements named.

1   Complete this table.

Compound name	The elements that make up this compound
zinc chloride	Zn *and* Cl
copper sulfate	Cu *and* S *and* O
zinc oxide	
copper nitrate	
zinc hydroxide	
magnesium chloride	
sodium nitrate	
iron sulfate	
	K *and* I
calcium carbonate	

2   A **chemical formula** tells you which atoms are involved, and how many of each.

The *letters* show which elements make up the compound. Metal atoms are written first.

$$Cu\ SO_4 \qquad MgCl_2$$

The *numbers underneath* show how many of each kind of atom make up one molecule of the compound. If no number is given, it means there is a single atom of that kind.

Another way of showing this kind of information is with **particle pictures**. The particle pictures drawn in the next table show the numbers of atoms, but not their exact arrangement or sizes.

ISBN: 9780170214650

Chemical Science

3 Complete this table. For particle pictures show C particles = black, Na = grey, Cl = yellow, Mg = silver, O = red.

Compound name	Chemical formula	Number of atoms	Particle picture
Carbon dioxide	CO_2	3	⚫⚫⚫
	NaCl	2	⚫⚫
Magnesium chloride			⚪⚪⚪
Magnesium oxide			
	$AlCl_3$		

16 Formula to name

Date for completion: / /

Parent sig: _____

Teacher sig: _____

SP1 Unit 3.7–3.9

Rule reminders:

- The name of a compound can tell you what elements make it up. In many cases, two element names are simply joined together and '-ide' put at the end.
- If the name ends in '-ate' or '-ite', this tells you that the compound contains oxygen atoms as well as the other elements named.
- A metal atom is written first, non-metal(s) last.
- Some atoms tend to remain in groups. See the following table for details.

Group formula	Name of this group
OH	hydroxide
SO_4	sulfate
NO_3	nitrate
CO_3	carbonate
HCO_3	hydrogen carbonate (or bicarbonate)
NH_4	ammonium

Using the above rules, write a name for each of the following 10 compounds. When counting the number of atoms, a number after a bracket shows that there are two (or three) groups of all the atoms inside the brackets. If no number is given, it means there is a single atom (or group) of that kind.

Chemical formula	Total number of atoms	Name of compound
LiCl	2	
$CaCO_3$	5	
K_2CO_3		potassium carbonate
NaCl		
MgO		
$CaSO_4$		calcium sulfate
$Al_2(SO_4)_3$	17	
$Cu(NO_3)_2$		
KOH		_____ hydroxide
PbI_2		lead _____

SP1 Unit 3.7–3.9

1 Write the formula for the following 10 compounds. For these particular compounds, no brackets are needed.

Name of compound	Chemical formula
Potassium chloride	
Magnesium carbonate	
Sodium hydroxide	
Calcium carbonate	$CaCO_3$
Calcium oxide	
Zinc sulfate	
Sodium hydrogen carbonate	$NaHCO_3$
Lithium nitrate	
Potassium iodide	
Zinc oxide	

2 This is a classification activity. Choosing from among the formulae named in activities 15, 16 and 17, make a list of all the compounds that fit into each of the following categories. In each case, you need only write the formula.

Two compounds with chlorine

_____

Two compounds with hydroxide

_____

Three compounds with carbonate

_____

One compound with hydrogen carbonate

_____

Three compounds with nitrate

_____

Three compounds with sulfate

_____

Chemical Science

ISBN: 9780170214650

18 Chemical or physical?

Date for completion: / /

Parent sig: _____
Teacher sig: _____

In a **chemical reaction** such as burning, atoms 'change partners' and become bonded to other atoms in new arrangements. In a **physical change** a substance may change state, or it may become mixed with other substances. We can't see atoms, but there are simple ways of checking whether the change you are seeing is a chemical reaction or a physical change:

- if it is **easily reversed** using physical methods, it is probably a **physical change**
- if it is **not easily reversed**, it is probably a **chemical reaction**
- if it **changes colour**, it is probably a **chemical reaction**
- if it **gives off a gas**, it is probably a **chemical reaction**
- if it **gives off heat**, it is probably a **chemical reaction**.

1   Here is a list of 12 different changes. Decide whether each is a physical change or a change that involves chemical reactions, and write it in the column where it belongs.

ice melting	*cooking an egg*	*dissolving sugar in water*
dissolving grease with detergent	*iron slowly rusting*	*fireworks exploding*
paint drying	*food scraps decaying*	*water boiling*
wood burning	*steam condensing*	*separating sand and water*

A physical change	Involves chemical reactions

Not sure:

2   Chemical reactions and physical changes are in some ways similar, and some ways different. Write letters A to H in the segment of the Venn diagram where each belongs.

A   may involve giving off a gas

B   may involve using heat

C   often involves heat being given off

D   not easily reversed using physical methods

E   usually possible to reverse it using physical methods

F   includes dissolving, mixing, changes of state

G   includes burning, exploding

H   always involves more than one substance

Chemical Science

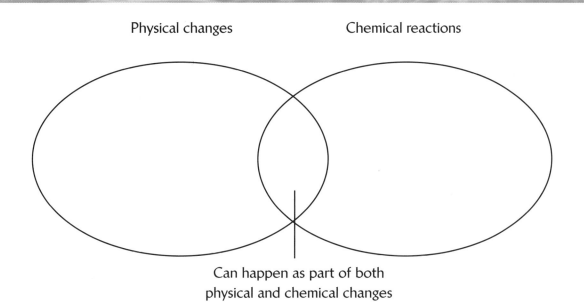

Physical changes   Chemical reactions

Can happen as part of both
physical and chemical changes

19   Introducing chemical reactions

Date for completion: / /

Parent sig: _____

Teacher sig: _____

SP1 Unit 3.8

During most kinds of chemical reaction, atoms change partners. Any chemical reaction can be summed up by an equation – either by a **word equation** or a **chemical formula equation**. In this workbook we start with word equations. It may help you to visualise 'particle pictures' as another type of equation. Any chemical equation must show the **reactants** (chemicals present at the start) and the **products** (chemicals at the end).

reactants ⟶ products

Complete the equations for reactions 2 to 6 below. In some cases heat energy is mentioned, but it is not necessary to do this in every equation. Best use colour-coding for particle pictures.

Reaction 1

If you heat a lump of carbon it will react with oxygen in the air. Carbon now exists in the form of carbon dioxide. We can show this reaction in three different ways:

Word equation:

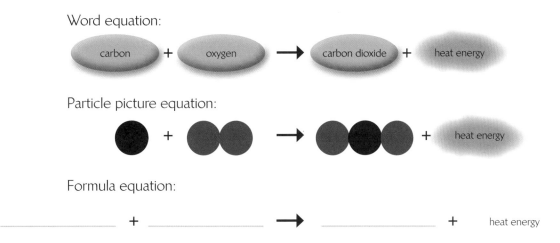

carbon + oxygen ⟶ carbon dioxide + heat energy

Particle picture equation:

● + ●● ⟶ ●●● + heat energy

Formula equation:

_____ + _____ ⟶ _____ + heat energy

ISBN: 9780170214650

Reaction 2

If powdered iron and sulfur are mixed and then heated together, these two elements will combine to form a compound.

Word equation:

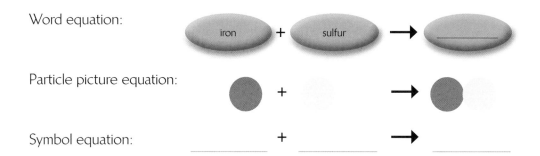

Particle picture equation:

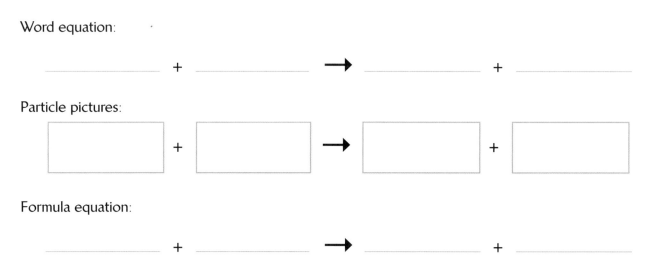

Symbol equation: _____ + _____ ⟶ _____

Reaction 3

Sodium hydroxide reacts with hydrogen chloride (a.k.a. hydrochloric acid), resulting in sodium chloride and water.

Word equation:

_____ + _____ ⟶ _____ + _____

Particle pictures:

+   ⟶   +

Formula equation:

_____ + _____ ⟶ _____ + _____

Reaction 4

Magnesium reacts with hydrochloric acid to produce magnesium chloride and hydrogen gas.

Word equation:

_____ + _____ ⟶ _____ + _____

Formula equation:

_____ + _____ ⟶ _____ + _____

CHALLENGE

Magnesium chloride is $MgCl_2$ and hydrogen is H_2. To balance the equation so that it correctly shows an equal number of each kind of atom before and after the reaction, you need to write '2' in front of one of the formulae. Decide which one.

Reaction 5

Calcium carbonate reacts with hydrochloric acid, producing CO_2 and $CaCl_2$ and H_2O.

Word equation:

_____ + _____ ⟶ _____ + _____ + _____

Formula equation:

_____ + _____ ⟶ _____ + _____ + _____

You will need to balance this equation by writing write '2' in front of one of the formulae. Decide which one.

Reaction 6

Word equation:

_____ + _____ ⟶ _____ + _____

Formula equation:

$2NaOH$ + $CuSO_4$ ⟶ Na_2SO_4 + $Cu(OH)_2$

20 Separation methods

Date for completion: / /

Parent sig: _____
Teacher sig: _____

SP1 Unit 3.10/3.11

Chemical reactions often produce a mixture of substances in the same container. We may need to separate these products. Several types of science equipment are used to get pure substances from mixtures.

1 Name each labelled feature of equipment below indicated by a ⟨?⟩ Also, name separation methods that A to F are generally used for.

A

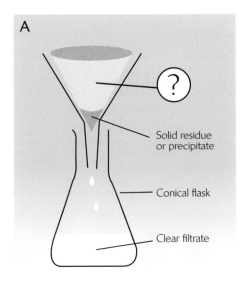

Solid residue or precipitate

Conical flask

Clear filtrate

B

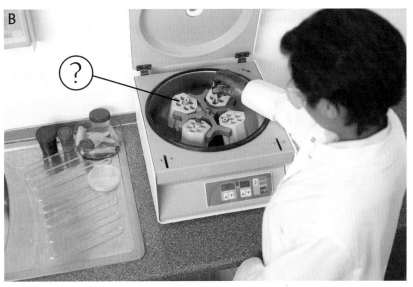

_____   _____

ISBN: 9780170214650

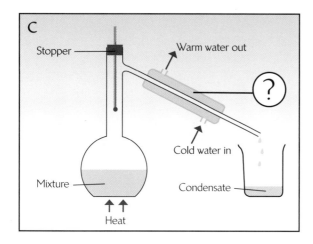

C

Stopper
Warm water out
?
Cold water in
Mixture
Condensate
Heat

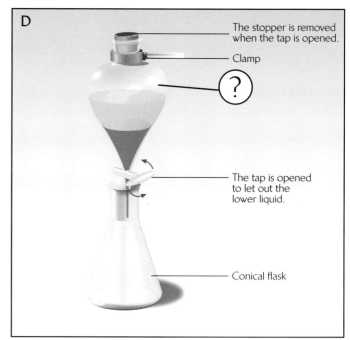

D

The stopper is removed when the tap is opened.
Clamp
?
The tap is opened to let out the lower liquid.
Conical flask

E

?

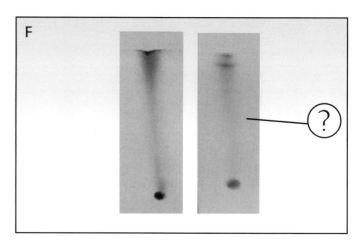

F

?

2 Which method and which equipment (from A, B, C, D, E, F above) would you use to separate each of the following mixtures? If none of the methods shown above is suitable, say so.

- oil and water

- alcohol and water

- fine sand and water

- fine sand and coarse sand

- pure water from sea water

- salt from sea water

- different dyes from ink

Date for completion: /   /

Parent sig: _____

Teacher sig: _____

SP1 Unit 3.10/3.11

Your little brother wants to make some popcorn, but ends up with a messy mixture of corn, salt and cooking oil, all mixed together in a big bowl. Instead of throwing everything away, you try to separate the mixture into three clean, dry separate lots of the three components. You have ordinary kitchen containers available. Decide what your methods of separating substances will be. Write your plan in bullet point form. List all the containers you will need. Also make a neat drawing of your set-up for at least one of the stages.

Chemical Science

ISBN: 9780170214650

22 Lab separation

Date for completion: / /

Parent sig: _____
Teacher sig: _____

SP1 Unit 3.10/3.11

Your teacher gives you a 10 gram mixture of dry sand + salt + iron sand. Only your teacher knows the exact proportions.

Your aim: To separate the mixture and find the exact proportions. Plan how to separate these into three separate piles without losing any. Decide what your methods of separating substances will be. Write your plan in numbered bullet point form. Also make a neat drawing of each stage, naming any science gear used. (The science room is equipped with all the gear listed in the 'Separation methods' activity on page 79–80.)

ISBN: 9780170214650

23 Inside the atom

Date for completion: / /   Parent sig: _____   Teacher sig: _____

SP1 Unit 3.12

1 The number of protons in the nucleus is the same as the atomic number of the element. Any proton has a mass of _____, and an electrical charge of _____. Any neutron, also in the nucleus, has a mass of _____, and _____ electrical charge. Electrons, which orbit the nucleus, have an insignificant mass and a charge of _____.

Complete the above by writing *1, +1, -1*, or *zero* in the correct places.

2 The diagram below represents one atom of a particular element. Which element? _____.
What is the atomic number of this element? _____ What is its mass number? _____

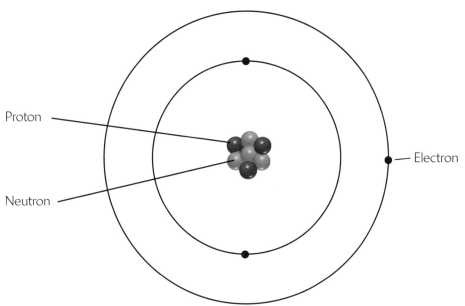

Proton

Neutron

Electron

3 Complete this table.

Element	Atomic Number	Mass Number	Number of		
			Protons	Neutrons	Electrons
Cl	17	35	17	18	7
Na	11	23			
O	8	16			
Mg		24			12
Ar	18			22	
Li			3	4	
Al			13	14	

Chemical Science

ISBN: 9780170214650

Will two elements react quickly, slowly, or not at all? It depends on their electron arrangements – especially the outermost layers. According to theory, electrons are arranged in a number of orbits around the nucleus. These electrons move at high speeds and follow rules that govern the number of electrons in orbit. These are the electron rules for the first 20 elements:

- the first orbit can hold two electrons maximum
- the second orbit can hold eight electrons maximum
- the third orbit can hold eight electrons maximum
- when an orbit is more than half full, electrons are arranged in pairs
- electrons take up the innermost position where there is a vacancy.

Example: Na, element 11, has 11 electrons. The electron arrangements can be drawn in a 'target' diagram like this:

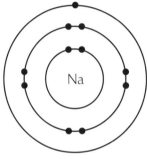

1   Draw similar diagrams for: C (atomic number 6), O (8), Mg (12).

2   Electron arrangements can be shown more simply. For example, Na with its 11 electrons can be also written **Na 2,8,1**. For each of the following atoms, write the electron arrangement in this simple way.

Element	Atomic number	Electron arrangement
Li	3	
O	8	
Ne	10	
Mg	12	
Al	13	
Cl	17	
Ar	18	
K	19	
Ca	20	

ISBN: 9780170214650

25 What's in the air?

Date for completion: / /   Parent sig: _____   Teacher sig: _____

SP1 Unit 3.14

Air is not one substance. As well as having dust particles and floating pollen, air is a complex mixture of gases, as this table shows. Water vapour is not listed. Reason: the percentage (%) of water vapour varies from about 0.1% up to about 2%, so it is simpler to give the numbers for 'dry' air.

Chemical name (alphabetical)	Chemical symbol	Average amount in the air
Argon	Ar	0.9 %
Carbon dioxide	CO_2	385 parts per million (ppm)
Helium	He	5 parts per million
Hydrogen	H_2	5 parts per 10 million
Krypton	Kr	100 parts per million
Neon	He	20 parts per million
Nitrogen	N_2	78 %
Oxygen	O_2	21 %
Ozone	O_3	1 part per 100 million

1   List the above nine gases in order from the biggest to smallest percentage component of air.

_____

_____

2   Use the information in the table to create a coloured pie chart of air's makeup, showing and labelling only three segments: **nitrogen**, **oxygen**, and **all other gases**. (Hint: 10% of the air = 36° of the pie. Use a protractor.)

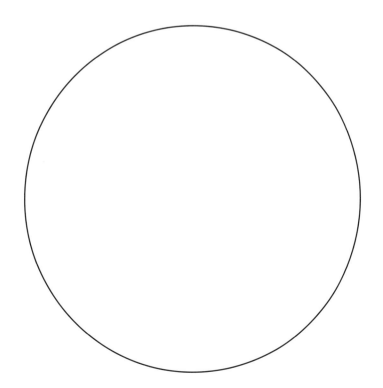

ISBN: 9780170214650

3 The picture below represents eight of the nine gases named in the table. On the diagram itself, label each of the 15 atoms and also name the five molecules. The colours are not real. To help you work out which is which, atoms are drawn to show their comparative sizes in proportion. A periodic table will give you an idea of which atoms are biggest and which are the smallest. (Oxygen atoms are shown in red here.)

4 Suggest a reason why the amount of oxygen in the air does not change much from one part of the world to another.

_____

_____

_____

_____

26 Hydrogen

Date for completion: / /

Parent sig: _____

Teacher sig: _____

SP1 Unit 3.15

1 Complete the following sentences. The first letter of each missing word is provided.

Hydrogen has the smallest atoms of all, with an a_____ mass of 1. It is also the most

abundant element in the u_____ . On earth, most hydrogen exists in w_____ ,

as H_2O. 'Free' hydrogen mainly exists in atom pairs, as H_2 m_____ . Hydrogen gas is only one

part per million of the a_____ , yet is the commonest element in your body, in

w_____ , proteins, c_____ and fat molecules. 'Free' hydrogen is

highly e_____ , so great care is needed when storing it. When hydrogen burns, it combines

with o_____ to form w_____ .

2 The drawing shows one way of making and collecting hydrogen gas.

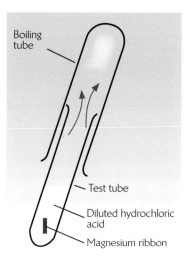

Boiling tube

Test tube

Diluted hydrochloric acid

Magnesium ribbon

Complete this word equation for the reaction:

_____ + _____ ➡ magnesium chloride + _____

Write this reaction as a formula equation:

_____ + HCl ➡ MgCl$_2$ + _____

To balance the equation so that it correctly shows an equal number of each kind of atom before and after the reaction, you need to write '2' in front of one of the above formulae. Decide which one.

3 This drawing shows the well-known 'pop' test for hydrogen gas.

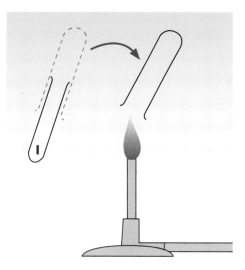

Explain why the boiling tube needs to be held upside-down.

_____

_____

Describe two things you are likely to see in the boiling tube at the moment of the 'pop' explosion. (Note: 'boiling tube' is the name for a wide-diameter 50 mL test tube.)

_____

_____

_____

Chemical Science

ISBN: 9780170214650

Date for completion: / /

Parent sig: _____
Teacher sig: _____

Chemical Science

1   Complete the following sentences. The first letter of each missing word is provided.

Oxygen is all around us in one form or another. It forms a major part of the earth's c_____

in the compound silicon dioxide (silica), the main component of r_____. It is also in

compounds like w_____, and of course it makes up 21% of the a_____.

Oxygen in the air is mostly in atom pairs, forming m_____ of O_2. The gas O_3, also known as

o_____, is a minor component of a_____, but is important because it screens

out most of the dangerous UV r_____. The gas oxygen (as O_2) does not actually

b_____, but is an essential part of any burning r_____.

Oxygen   Ozone

This drawing shows one set-up used to make and connect oxygen gas in a science room.

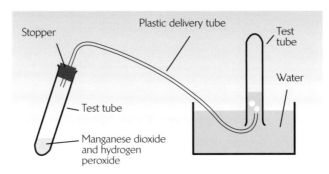

Stopper

Plastic delivery tube

Test tube

Water

Test tube

Manganese dioxide and hydrogen peroxide

2   This method is known as 'collecting gas by downward displacement of water'. Explain the purpose of this technique, instead of just letting the gas go straight into a test tube.

_____

_____

_____

CHALLENGE

3   The manganese dioxide is a **catalyst**, and is not used up in the reaction. Explain what a catalyst does.

_____

_____

_____

ISBN: 9780170214650

4 Complete these equations of the reaction for making oxygen:

Word equation:

hydrogen peroxide $\longrightarrow$ water + _____

Formula equation:

H_2O_2 $\longrightarrow$ _____ + _____

To balance this equation so that it correctly shows an equal number of H and O atoms before and after the reaction, you need to write '2' in front of two of the above formulae. Decide which two.

5 This drawing shows a traditional test to find whether a gas is oxygen or not. If the gas is oxygen, describe what will happen.

_____

_____

_____

_____

_____

Glowing splint — Boiling tube

Gas

6 Most living things use oxygen to release energy from glucose sugar. This process supplies energy to all living cells, and is usually given the name **cell respiration**. Complete this simplified word equation for cell respiration.

glucose + _____ $\longrightarrow$ _____ + _____ + energy

---

28 Carbon dioxide

Date for completion: / /

Parent sig: _____

Teacher sig: _____

SP1 Unit 3.17

1 Complete the following sentences (the first letter of each missing word is provided), then use these in the puzzle on the following page.

Carbon dioxide is essential to life on Earth, because g_____ plants feed on it and use it in the

process of p_____. CO_2 is also a major g_____ gas, one of several

gases that keep the planet warmer. However, steady i_____ in atmospheric CO_2

seems to be pushing the Earth's average t_____ to higher levels. CO_2 gas presently

makes up about 0.038% of the air we b_____ in. The air that you e_____ is about

4% CO_2. Solid CO_2 is also known as d_____ ice, often used to give special-effects 'smoke'.

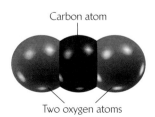

Carbon atom

Two oxygen atoms

ISBN: 9780170214650

2 It is easy to make CO_2: react any acid (like hydrochloric acid, HCl) with any carbonate chemical. Complete the following equations.

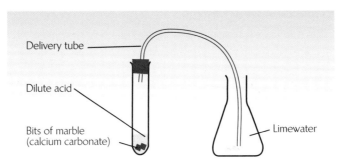

Word equation:

hydrochloric acid + _____ ➜ _____ + _____ + water

Formula equation:

_____ + $CaCO_3$ ➜ $CaCl_2$ + _____ + _____

CHALLENGE

To balance the equation so that it correctly shows an equal number of each kind of atom before and after the reaction, you need to write '2' in front of one of the above formulae. Decide which one.

3 The simplest way of checking whether a gas is CO_2 (or not) is to bubble it through **limewater**. Limewater is a dilute solution of **calcium hydroxide**. Even a small amount of CO_2 will soon make the limewater look 'milky', by producing tiny amounts of calcium carbonate. Complete the following equations, which sum up the limewater test.

Word equation:

_____ + calcium hydroxide ➜ calcium carbonate + water

Formula equation:

_____ + $Ca(OH)_2$ ➜ _____ + _____

4 Sarah has made a beaker full of what is supposed to be CO_2, and her teacher asks her to check whether it can put out a small flame. Make a drawing to show what Sarah could do to check whether the flame-test works. Describe what she is likely to see.

_____

_____

_____

_____

_____

_____

_____

_____

_____

5 Sarah makes another beaker of what is supposed to be CO_2, and now wants to check whether the gas she has made is denser than air. Make a list of practical steps and drawings that advise Sarah what to do to check the density of the gas.

_____

_____

_____

_____

_____

_____

_____

_____

29 Chemical round-up

Date for completion: / /

Parent sig: _____

Teacher sig: _____

SP1 Unit 3.18

Do this task by writing the matching terms in the right hand column. Here are seven of the 10 you will need (the remaining three are for you to puzzle out): _electrons; molecule; organic; alloy; concentration; element; boiling point._

	Description	Matching word(s)
1	The temperature above which a substance cannot exist as a liquid.	
2	A substance that has only one single type of atom.	
3	A **substance** that has two or more kinds of atom chemically combined with each other.	
4	A **particle** that has two or more atoms that are chemically combined with each other.	
5	A substance with two or more components that are not chemically combined with each other.	

Continued over

ISBN: 9780170214650

	Description	Matching word(s)
6	A mixture of metals, such as bronze.	
7	The number of particles in each unit volume.	
8	The negative subatomic particles in atoms.	
9	An arrangement based on the number of protons in each atom.	
10	A word for carbon containing compounds like plastics and sugars.	

30   Nitrogen

Date for completion: /   /

Parent sig: _____
Teacher sig: _____

SP1 Unit 3.18

1   Complete the sentences below by writing in items chosen from this list: *fixing, legume, 78, enrich, protein.*

Nitrogen gas (N_2) makes up _____ % of the air around us. All living things need the element nitrogen, mainly in the _____ substances in their bodies. However, plants and animals are not able to use N_2 gas directly. But nitrogen - _____ bacteria are able to directly use N_2 to make protein. Clover and other _____ plants have nodules attached to their roots, with these bacteria living inside. For this reason, plants such as clover are an important part of good farming practice, as they _____ the soil.

A molecule of N_2 gas.

Nodules containing nitrogen-fixing bacteria on the roots of a legume plant.

CHALLENGE

2   Artificial fertilisers contain several of the approximately 18 elements that plants need. Use the figures on the bag at right to calculate how much each of the following elements are in 10 kg of the bag's contents. Give all weights in gram amounts (*ppm* is short for 'parts per million'.)

Nitrogen _____

Phosphorus _____

Potassium _____

Magnesium _____

Iron _____

Boron _____

Zinc _____

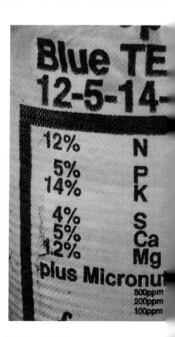

Blue TE
12-5-14-

12%	N
5%	P
14%	K
4%	S
5%	Ca
12%	Mg

plus Micronu

500ppm
200ppm
100ppm

ISBN: 9780170214650

Date for completion: / /
Parent sig: _____
Teacher sig: _____

SP1 Unit 3.19/3.20

Chemical Science

Reaction 1

A fire is a fast chemical reaction between a fuel and oxygen. In most cases, 'fuel' means carbon compounds such as those in wood, natural gas, petrol or diesel. When fire has all the oxygen supply it can use, the result is **complete combustion**, and CO_2 and H_2O are the only chemical products.

Reaction 2

If a fire is partly starved of oxygen it switches to different reactions known as **incomplete combustion**. To begin with, the reaction below produces the highly dangerous gas carbon monoxide, **CO**.

Reaction 3

If the fire is further starved of oxygen, it won't go out immediately; it will switch to another **incomplete combustion** reaction. The one below will produce unburned carbon (**C**), and not CO or CO_2.

A Write a formula equation for each of reactions 1, 2 and 3, for situations in which the fuel is natural gas, also known as **methane**, formula CH_4. To begin with, each equation you write will not be balanced. To balance each equation so that it correctly shows an equal number of each kind of atom before and after the reaction, you need to write '2' or '1.5' in front of some of the formulae. Decide which ones.

Reaction 1 (complete combustion)

$$CH_4 + O_2 \longrightarrow \underline{\hspace{3cm}} + H_2O$$

Reaction 2 (incomplete combustion)

$$CH_4 + O_2 \longrightarrow \underline{\hspace{3cm}} + H_2O$$

Reaction 3 (incomplete combustion)

$$CH_4 + O_2 \longrightarrow \underline{\hspace{3cm}} + H_2O$$

B Carbon monoxide is a highly poisonous gas. Describe the ways it affects the blood.

_____

_____

_____

ISBN: 9780170214650

C Explain what causes an incomplete combustion flame to have an orange-yellow colour.

_____

_____

_____

_____

_____

_____

_____

_____

Yellow or orange flames like this are caused by millions of glowing hot specks of carbon. Sooty black smoke is unburnt carbon, showing that the fire is being starved of oxygen.

32 Explosives

Date for completion: / /

Parent sig: _____

Teacher sig: _____

SP1 Unit 3.21–3.23

Gunpowder was invented more than 1000 years ago by a group of Chinese, then independently by Europeans in the 1400s. Like any explosive, gunpowder works by producing a sudden rush of hot gases. But gunpowder's chemical reactions are not particularly fast: high explosives like TNT react much faster. Gunpowder's main modern use is in fireworks.

Modern high explosives were first developed by Alfred Nobel of Sweden. At the age of 17 he heard of a new colourless liquid, nitro-glycerin, a recently-invented explosive. But there was a problem; the slightest shock could cause it to explode without warning. After years of trials, Alfred discovered that nitro-glycerin could be made safer by absorbing it in clay. In 1866, he patented his mixture and gave it a new name: dynamite.

Alfred Nobel hoped that dynamite would be used for engineering work only, but it soon became used in warfare. His brother Emil was killed in the family dynamite factory. Alfred made a fortune through his many chemical inventions, and gave his great wealth to create the Nobel Prize system.

Tri-nitro-toluene (TNT), a yellow solid, is a powerful explosive. Like dynamite and other modern explosives, it is not set off by a burning fuse. A small explosive charge called a detonator provides the 'trigger' that jolts it to react.

One kilogram of TNT contains 4 million joules of chemical energy, less than the energy in one kilogram of petrol. The difference is that TNT can convert its chemical energy to heat energy in a fraction of a second.

The explosions you see in movies are made with petrol and propane mixed together so they burn with a colourful flame and lots of smoke. The orange flame colour is given by hot bits of carbon. An explosion of dynamite or TNT is far more dangerous but doesn't look as dramatic. A TNT explosion gives a brief flash of light, a cloud of dust, and a shock wave strong enough to tear solid objects to pieces.

ISBN: 9780170214650

1 Underline the main points in the story on the previous page.

2 Write a summary of the main facts in the above story in 10 to 15 bullet points.

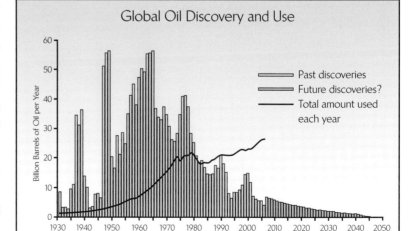

33 Fuels and the future

Date for completion: / /

Parent sig: _____

Teacher sig: _____

SP1 Unit 3.24

For the past hundred years, modern societies have depended on petroleum oil as a cheap source of energy. From the 1960s to 2000, demand rose to more than 12 million barrels a day. New crude oil discoveries have become smaller and the future for liquid-energy supplies is very uncertain.

The graph (right) is based on figures from the oil industry, and is correct up to 2006. Use the graph to answer the following questions.

Global Oil Discovery and Use

Billion Barrels of Oil per Year

■ Past discoveries
■ Future discoveries?
— Total amount used each year

ISBN: 9780170214650

1   Give the dates of the five years when the biggest oil discoveries took place.

_____

2   Describe the overall trend of new discoveries in the early 1960s.

_____

_____

3   On the basis of the graph, predict how likely it will be that big future discoveries will be made, as big as those in the 1960s.

_____

_____

4   Describe the overall trend of amounts of oil used since 1930.

_____

_____

5   From a newspaper or the internet, find the present price of crude oil per barrel (bbl).

$_____. One bbl = 200 litres, so divide the figure by 200 to get the price of one litre of

crude _____

6   What is the present price of regular petrol at the pump? $_____

7   Suggest three reasons why the per litre price at the pump is much more than the price of crude oil.

_____

_____

_____

8   Renewables like wind energy and biofuels (at present) make up less than 10% of the world's energy supply. What do you predict will happen to energy supply over the next 40 years? Give reasons for your answer.

_____

_____

_____

_____

_____

_____

_____

_____

_____

_____

ISBN: 9780170214650

Physical Science

1   Physical units

Physical science relies on accurate measurements, and uses particular units. By 'unit' we mean a form of measurement, such as kilometres. We will use only **SI units** (short for 'Systeme Internationale'). There are also non-SI units, like calories, miles or hours.

1   Complete this summary of important SI units and their symbols.

Distance is measured in **metres**, symbol _____.

Time is measured in _____, symbol **s** (*not* 'sec').

Mass is measured in _____, symbol **kg**.

_____ is measured in _____, symbol **J**.

Force is measured in **newtons**, symbol _____.

Current is measured in **amps**, symbol _____.

In many cases the main SI unit is too small or too big, so a related unit is used, like millimetre or kilometre. These mostly differ by multiples of 1000, so when converting you need to move the decimal point in steps of three places. These units are always identified by their prefix, the first part of the name.

> **Important!**
> Whichever unit of measurement is used, you should always write the unit symbol as well as the number eg: length = 25 mm, *not* length = 25.

Number of times this unit is bigger or smaller than the standard SI unit	Prefix	Symbol
10^{12}	tera -	T
10^{9}	giga -	G
10^{6}	mega -	M
10^{3}	kilo -	k
10^{-3}	milli -	m
10^{-6}	micro -	μ
10^{-9}	nano -	n

2   Complete these measurement conversions. You need to decide where to move the decimal point, left or right, and how many places.

6432 mm, converted to metres (m)   = _____

17.2 km, converted to metres (m)   = _____

328 g, converted to kilograms (kg)   = _____

17 kJ *, converted to joules (J)   = _____

0.2 seconds, converted to milliseconds   = _____

0.2 seconds, converted to microseconds   = _____

3 gigabytes, converted to megabytes   = _____

1.5 litres, converted to millilitres (mL)   = _____

600 milliamps, converted to amps (A)   = _____

8 micro-metres **, converted to mm   = _____

0.000008 metres, converted to mm   = _____

760 nm ***, converted to mm   = _____

Remember to write the correct units!

* the amount of energy in 1 g of sugar.

** the diameter of a human red blood cell.

*** the wavelength of red light.

ISBN: 9780170214650

3 Round each of the following measures to three significant figures (3 SF). In rounding, you do not move the decimal point.

Original	Rounded to 3 SF	Original	Rounded to 3 SF
3792.1 g		0.007231 mm	
25.87 mm		2567.4 km	
25.01 kg		520.3333 g	

Biography

Maria Sklodowska was born in Poland in 1867, the daughter of a school teacher. Her homeland was then dominated by Russia. She became involved in a students' revolutionary organisation and eventually had to flee Poland as a political refugee. In 1891 she went to Paris to continue her studies in Physics. She met Pierre Curie in 1894 and in the following year they were married. She gained her Doctor of Science degree in 1903, and following the tragic death of Pierre Curie in 1906 took his place as Professor of Physics, the first time a woman had held this position.

Her early researches, together with her husband, were performed under difficult conditions, laboratory arrangements being poor. The discovery of radioactivity in 1896 inspired the Curies in their brilliant research, and analysis, which led to the discovery of radium and also polonium, named after the country of Marie's birth. Mme. Curie developed methods for the separation of radium in sufficient quantities to allow for the careful study of its properties, medical properties in particular, such as X-rays.

Mme. Curie throughout her life actively promoted the use of radium to alleviate suffering. During World War I, assisted by her daughter Iréne, she devoted herself to this remedial work. She retained her enthusiasm for science throughout her life and did much to establish a radioactivity laboratory in her hometown of Warsaw.

Mme. Curie, quiet and unassuming, was held in high esteem by scientists throughout the world. Her work included investigations on radioactive substances and radioactivity. The importance of her work is reflected in the numerous awards she was given. Together with her husband, she was awarded half of the Nobel Prize for Physics in 1903. In 1911 she received a second Nobel Prize, this time in Chemistry for recognition of her work in radioactivity. She was the first woman to receive a Nobel Prize, and remains the only woman to receive two.

The Curie's elder daughter, Iréne, married Frédéric Joliot and they were joint recipients of the Nobel Prize for Chemistry in 1935. They both took a lively interest in social problems, and as Director of the United Nations' Children's Fund Frédéric received on its behalf the Nobel Peace Prize in 1965. Mme. Curie died in France, after a short illness, on 4 July 1934.

ISBN: 9780170214650

1 Underline key facts in the above biography.

2 Write a summary of the main points in three paragraphs under the headings **Personal life**, **Scientific discoveries**, **Prizes and awards**. The total length should be 100 to 150 words.

_____

_____

_____

_____

_____

_____

_____

_____

_____

_____

_____

_____

_____

_____

_____

_____

_____

_____

_____

_____

_____

_____

_____

_____

_____

_____

_____

_____

_____

_____

Physical Science

ISBN: 9780170214650

We still don't know what energy actually is, but we do know a lot about its different forms and how to measure them. Energy occurs in a number of different forms, usually grouped into two main kinds.

Active forms	Potential forms
All of these involve movement.	All of these are 'stored'.
Kinetic energy	Chemical potential energy: Ep (chem)
Heat energy	Elastic potential energy: Ep (elas)
Light energy	Gravity potential energy: Ep (grav)
Sound energy	Nuclear potential energy
Electric energy	Magnetic potential energy

Each of the following devices and situations changes energy from one form to another. In each case, write the main energy input and the main 'energy output'.

Energy in: _____
Energy out: _____

Energy in: _____
Energy out: _____

Energy in: _____
Energy out: _____

Energy in: _____
Energy out: _____

Energy in: _____
Energy out: _____

Energy in: _____
Energy out: _____

Energy in: _____
Energy out: _____

Energy in: _____
Energy out: _____

Energy in: _____
Energy out: _____

Energy in: _____
Energy out: _____

Energy in: _____
Energy out: _____

Energy in: _____
Energy out: _____

Physical Science

Complete these sentences.

1   A can of diesel represents 'stored' _____ energy.

2   Energy in stretched springs is called _____ energy.

3   Energy in objects because of their high position is called _____ energy.

4   All forms of energy that are **stored** is called _____ energy.

5   All objects in motion possess _____ energy.

6   A battery changes _____ energy into _____ energy.

7   An electric blanket changes _____ energy into _____ energy.

8   As he falls, a skydiver's energy changes from _____ to _____ energy.

9   Kinetic energy is the energy of _____.

10   When you hit a tennis ball, the _____ energy of your arm is changed into kinetic energy of the ball.

Energy changes can often be simplified in an 'energy equation' which summarises the main overall energy changes. Example for a light bulb:

$$\text{electric E} \rightarrow \text{heat E} + \text{light E}$$

Write the main energy equations for each of the following situations.

11   An electric toaster.

_____

12   A skateboarder going down a ramp.

_____

13   A car moving at constant speed.

_____

14   A power drill.

_____

15   A computer.

_____

Physical Science

ISBN: 9780170214650

It may not be obvious, but this boulder contains a great amount of energy. It's not moving, it's not electrically charged, and it's not going to explode. However, it holds energy because of its position.

The energy that any object has because of its high position is known as **gravitational potential energy**, or **Ep (grav)** for short.

The amount of Ep (grav) of an object depends on two factors:

- Its **mass**. Double the mass means double the energy.
- Its **height**. Double the height means double the energy. This depends where you are measuring the height from. Usually it is taken as the distance above ground level, but you could also calculate Ep (grav) from floor level, even if the floor is many stories up.

1 Complete these sentences.

Whenever any object is lifted upwards it gains _____

_____ energy, often written _____ for short . The

_____ it is lifted upwards is a measure of the energy gained. If you lift a

heavy weight and allow it to fall, its energy is changed to _____ as it falls, and

then at the moment of impact to _____ and _____.

2 These five grey rocks have different amounts of gravitational potential energy, compared to the level they would fall to if nudged a little to the right. List them in order from the one with most Ep (grav) down to the least. The three big rocks are all the same mass, and each has exactly three times the mass of a small rock.

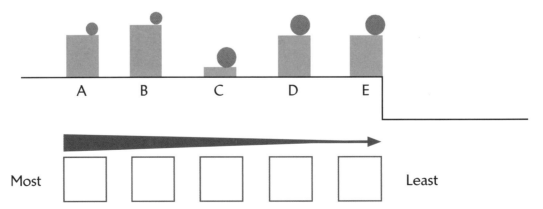

Most ☐ ☐ ☐ ☐ ☐ Least

Physical Science

3 Carlos tells you that most of New Zealand's electricity is 'hydro' and is made from water. Explain to him why he is not quite correct, and explain how hydroelectricity is actually made.

_____

_____

_____

_____

_____

6 Energy efficiency

Date for completion: / /

Parent sig: _____

Teacher sig: _____

SP1 Unit 4.5

> *The Law of Conservation of Energy:*
> **Energy can be changed from one form to another,
> but cannot be created and cannot be destroyed.**

This law is a universal fact of all energy changes. However, every energy conversion produces some heat – and in some cases lots of heat.

For example an 'old' light bulb:

electric energy ⟶ light energy (about 5%) + **heat energy** (about 95%).

The heat energy is not 'lost'. It is still there, warming the surroundings. But from a human point of view this heat is wasted, and we say that the bulb is only '5% efficient'.

1 Write energy equations for each of the following. In these equations E is used in place of the word 'energy'. In each case, circle the useful energy output(s), and underline the energy that is wasted as heat E or sound E.

Food blender:

_____ E ⟶ _____ E + _____ E + _____ E

Car engine:

_____ E ⟶ _____ E + _____ E + _____ E + _____ E

Candle flame:

_____ E ⟶ _____ E + _____ E

2 **Energy efficiency** is how much useful energy is obtained from a process, compared to how much is put in. Calculate the percentage efficiency using this formula:

$$\text{Efficiency} = \frac{\text{Useful energy}}{\text{Input energy}} \times 100$$

Physical Science

ISBN: 9780170214650

Calculate the efficiency in each of the following situations. Show your working. Round each answer to three significant figures.

Old type light bulb: 60 J of energy used per second, 3 J of light produced.

_____

_____

Fluorescent light bulb: 15 J of energy used per second, 3 J of light produced.

_____

_____

_____

Electric drill: 500 J used each second, 360 J of kinetic energy produced.

_____

_____

_____

Car engine: 750 kJ used in one minute, 500 kJ ends up as heat and noise.

_____

_____

_____

3 'Heat energy is always wasted energy.' Is this always true, false, or mostly true? Explain.

_____

_____

_____

_____

_____

_____

_____

_____

_____

| Date for completion: | / / | Parent sig: _____ |
| | | Teacher sig: _____ |

SP1 Unit 4.6

Annie and Luke were given the task of finding out how much chemical energy is contained in a peanut. They burnt one while aiming the flame under a boiling tube containing 20 mL of water. They already knew that it takes 4.2 J of energy to raise the temperature of 1 mL of water by 1 °C. They carried out the experiment and got the results shown in the table below as the water heated up over 150 seconds, when the nut-flame died.

Plot their results on the graph below using all graphing rules. Put Time on the X (horizontal) axis.

Time (s)	Temp (°C)
0	15
30	18
60	22
90	27
120	31
150	32

1 What was the temperature of the water at the beginning?_____.

And at the end of the experiment? _____

2 What was the temperature rise of the water at the time the nut burned out? _____

3 Calculate the number of joules of energy that the water received.
Multiply temperature rise x volume of water x 4.2. (It takes 4.2 J of heat to raise the temperature of 1 mL of water by 1°C.) Show all steps in your calculation.

_____

_____

_____

_____

Physical Science

ISBN: 9780170214650

4 What is the unit for energy? _____

CHALLENGE

5 Was all the energy from the burning peanut used to heat the water? How could this have affected their results? Explain.

_____

_____

_____

CHALLENGE

6 Suggest one practical improvement to the method used by Annie and Luke.

_____

_____

_____

8 Heat and temperature

Date for completion: / /   Parent sig: _____ Teacher sig: _____

SP1 Unit 4.7

Physical Science

1 Of these two containers (bucket and cup) of water, which has the highest temperature?_____

And which contains the greatest amount of heat? _____

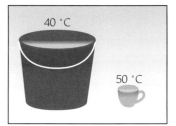

You have just proved to yourself that heat and temperature are not the same thing. Temperature is measured in degrees centigrade (°C). Heat is a **total quantity** of energy, measured in joules (J).

CHALLENGE

2 **Calculating heat**. (This task is for students confident of their maths ability.)

Heat is a total quantity of energy, and depends on temperature change and on the amount and type of substance.

The amount of heat gained (or lost) by water can be calculated using the formula below. Δ is a scientific symbol used to represent 'a change in the amount'.

$$\Delta H = m \times \Delta t \times 4.2$$

ΔH is the total heat increase or decrease in joules.

m   is the mass of water in grams.

Δt is the temperature rise (or fall) in °C.

4.2 is J/g/°C, because 4.2 J heat raises the temperature of 1 gram water by 1 °C.

Calculate the amount of kilojoules of energy involved in A and B. Show all your working. Use the correct units. Set your calculation out in the following three steps: FSC.

• Formula: write the formula.

• Substitute: with letters in place of numbers.

• Calculate final answer, and write the correct units.

A Your kettle holds 2000 grams of water. You fill it with 15 °C water from the tap. Calculate how much energy is needed to heat all the water to boiling point.

F: _____

S: _____

C: _____

B After your long shower, the electric hot water cylinder has to heat water from the intake temperature of 15 °C up to the set temperature of 60 °C. The amount of hot water used was 17 litres (17 kg). Calculate how much energy is needed.

F: _____

S: _____

C: _____

3 About kelvin. Normally, temperature is measured in degrees **Celsius** (symbol °C, also known as centigrade). But temperature can also be measured in **kelvin**, a SI unit. One kelvin (symbol K), is the same temperature measurement as one degree Celsius, but the kelvin scale starts at absolute zero, -273 °C. Convert the following temperatures to kelvin, K. (For technical reasons we write K and not °K. Saying 'degrees kelvin' is not correct.)

A freezing point of water   0 °C   =   _____

B boiling point of water   100 °C   =   _____

C freezing point of CO_2 -78 °C   =   _____

D melting point of lead   327 °C   =   _____

E absolute zero   -273 °C   =   _____

9

Heat gain and loss

Date for completion: / /

Parent sig: _____

Teacher sig: _____

SP1 Unit 4.7–4.9

Heat energy moves from warmer to colder things in three ways:
convection, conduction and **radiation**.

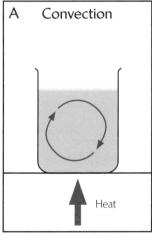

A   Convection

Heat

This involves the circulation of liquids and gases. Hot regions rise, cool dense regions sink.

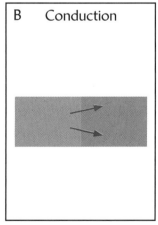

B   Conduction

This is the transfer of heat from a hot material to a less hot material, but only if they are in contact. Metals conduct heat rapidly.

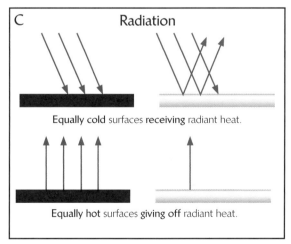

C   Radiation

Equally cold surfaces **receiving** radiant heat.

Equally hot surfaces **giving off** radiant heat.

Can send heat through still air or empty space. This drawing shows the importance of colour in absorbing and giving off IR (infrared) heat. Texture is also important: shiny surfaces absorb and give off least heat, while matt (rough) surfaces absorb and give off most heat.

ISBN: 9780170214650

1 Explain why in sunny weather it's more practical to wear light-coloured clothing, compared to dark colours.

_____

_____

_____

2 Explain why kettles are usually white or shiny, not dark colours.

_____

_____

3 Explain why solar heating panels are painted black.

_____

_____

4 Explain why in cold weather a person with light-coloured clothing will lose heat more slowly than someone wearing dark colours, if conditions are the same in other ways.

_____

_____

5 Of the three methods of heat transfer (**conduction**, **convection**, **radiation**), which best fits each of the following descriptions:

• is able to send energy through empty space. _____

• only happens if there is physical contact. _____

• only happens in gases and liquids. _____

• happens effectively in metals. _____

• travels at the speed of light. _____

6 Explain why layers of newspaper are good at keeping fish and chips hot, and yet also good at keeping ice cream cold.

_____

_____

_____

7 When you stand on a stone floor it feels colder than carpet, even though both surfaces are actually at the same temperature. Explain why the stone feels colder.

_____

_____

_____

8 Does the white colour of polar bears help keep them colder or warmer, compared to brown (grizzly) bears in the same conditions? Explain (A) for sunny conditions, and (B) for night conditions.

A_____

_____

B_____

_____

Date for completion: / /   Parent sig: _____
Teacher sig: _____

SP1 Unit 4.8–4.9

Heat is a form of energy. We use it for a lot of things, like warming our homes and cooking our food. Heat always tends to move from warmer to cooler places, and it does so in three ways: conduction, convection and radiation.

Pan handle is an insulator and doesn't **conduct** heat very well.

Soup is heated in the pan by **convection**. The hot soup rises. Cool soup sinks to take the hot soup's place.

Heat energy from the stove is transferred to the pan by **conduction**.

Use the following words to fill in the blanks: *gases, cooler, conductors, insulators, liquids, currents, energy, metal, hot, plastic, temperature.*

Conduction occurs when heat _____ (1) is passed directly from

one place to another. When you stir a pot of soup on the stove with a metal spoon,

the spoon's _____ (2) will rise. The heat is being conducted from the

_____ (3) soup to the _____ (4) spoon. Metals are excellent

_____ (5) of heat energy. Wood and plastics are not. These 'poor' conductors

are called _____ (6). That's why pans are made of _____ (7),

while the handle is made of a strong _____ (8). Convection is the movement of

_____ or _____ (9) from a cooler area to a warmer area. In a

glass container, you can see these circular convection _____ (10) . Warmer soup

moves up from the hot area at the bottom of the pan to the top where it is cooler. Cooler soup

then moves to take the warmer soup's place.

Physical Science

ISBN: 9780170214650

Date for completion: / /   Parent sig: _____
Teacher sig: _____

SP1 Unit 4.4–4.9

The following key words below are hidden in this review word puzzle. Circle the key words in the puzzle as you find them. A score of 8 is good; 12 is excellent; 15 is amazing!

Key words: CAR, CHEMICAL, EFFICIENCY, ELASTIC, WIND, ELECTRICAL, ENERGY, FORM, FUEL, GRAVITATIONAL, HEAT, JOULE, KINETIC, NUCLEAR, POTENTIAL, OIL, NONRENEWABLE, RADIO, SOLAR, SOUND, WATT, TRANSFORM.

```
M R O F C H E M I C A L T O C A R
E L D N I W I J J A A T S Y R D A
L K H G I L S O I I G T C C F O D
B O I D E F U F T T A W X N Z E I
A D O N N L G N E R O O B E M E O
W A N R E D E M X O A V E I D O S
E W B U C T K C A T Y N F C S M R
N E X A O D I Z T M E O S I N D A
E G H P H S O C I R N T H F U E L
R C N H O W E L G H I R T F O X O
N U M E R L G Y A N U C L E A R S
O G R A V I T A T I O N A L O A M
N I A T R O D I C I T S A L E N T
```

Date for completion: / /   Parent sig: _____
Teacher sig: _____

SP1 Unit 4.4–4.9

Use some of the key words contained in the wordfinder word list (above) to complete the sentences below. Use each word only once.

Scientists call stored energy _____ energy. Three forms of potential energy are:

- _____ (found in food and _____ like coal and _____).

- _____ (found in objects up high).

- _____ (found in springs).

Most of our chemical potential energy (like oil, coal) is non-_____, and will run out one day.

Sources like _____, waves and _____ are being developed.

Physical Science

_____ can change from one _____ into another. In a _____, chemical energy is changed to _____ energy and heat. In a radio, electrical energy is changed to _____. In a washing machine, _____ energy is transformed to kinetic energy and _____. A nuclear power station changes energy into electrical energy. If a machine turns most of the energy it uses into useful energy, it is said to have high _____. Energy is measured in _____ units.

Date for completion: / /
Parent sig: _____
Teacher sig: _____

SP1 Unit 4.11

1 The first one or two letters of each missing word is given. Write the whole word in each gap.

Almost all the world's electric energy is generated by moving co_____ of wire past ma_____ fields. Result: the movement causes an electric c_____ to flow. In the overall process, ki_____ energy is changed to e_____ energy. In all these cases, there has to be a source of energy to drive the generators. In New Zealand, most electricity generators are driven by steam pressure, w_____ , or the pressure of falling w_____. Any power station that relies on steam is described as t_____. The original source of heat energy to make steam is mostly coal and other f_____ fuels. Places like Waireki and Ohaaki are able to use g_____ heat for electricity generation.

2 Compare and contrast electric motors and internal combustion engines (ICEs). Suggest three or more advantages of electric motors, compared to ICEs:

_____

_____

_____

Suggest one advantage of ICEs (engines using petrol and diesel), compared to electric:

_____

3 A car has several different devices that use electric motors. List at least four.

_____

_____

_____

CHALLENGE

Physical Science

ISBN: 9780170214650

4 This table lists different kinds of power station used to generate electricity. List at least one main advantage and at least one main disadvantage of each type of generation. State each advantage/ disadvantage in five words or less.

Type of power station	Advantage	Disadvantage
Runs on coal, oil or natural gas		
Uses heat from nuclear reactions		
'Hydro': powered by falling water		
Solar powered		
Wind powered		
Geothermal		

14 Energy collection

Date for completion: / /

Parent sig: _____

Teacher sig: _____

Write the following words in the correct places in statements 1 to 8. Some words will be used more than once: *chemical; photosynthesis; gravity; 'hydro'; petrol; friction; light; diesel; kinetic; burned; heat.*

1 Electrical energy is particularly useful to us because it can be easily changed into other forms of energy

that we can use, such as _____, and _____.

2 The main source of electrical energy in New Zealand is _____, with

_____ in second place.

3 The electrical energy from a battery comes from _____ energy provided by reactions inside the battery.

4 _____ energy from the sun is changed by _____ in plant leaves

into _____ potential energy in sugars and starch. The cells of our body change

_____ potential energy in food into heat energy.

5 Fuels such as _____ and _____ contain chemical potential energy.

This is changed into heat energy when they are _____.

6 When a car is moving, some of its _____ energy is changed into heat energy by the

force of _____.

Physical Science

7 When you walk and run, _____ potential energy in the food you have eaten is changed by muscles to _____ energy and _____ energy.

8 When a skydiver steps out of the plane and begins to fall, some of her _____ potential energy in changed into _____ energy

Uninhabited Kohiwai Point, illustrated below, has been chosen as the site for a new settlement of 20,000 people, plus manufacturing industries. Add to the drawing your own ideas and details of a range of developments which can provide energy for the area. Some of the energy sources are visible in the picture, some may not be. Think also about energy distribution around the area. Next to each feature you draw add 'SR' if it will be sustainable and renewable, or 'NS' if you think it will be non-sustainable and non-renewable in the long run.

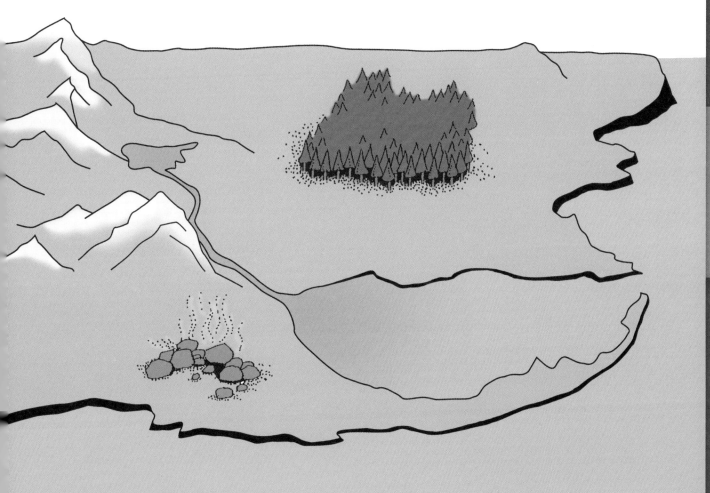

ISBN: 9780170214650

16 E-M radiation

SP1 Unit 4.12

Date for completion: / /

Parent sig: _____
Teacher sig: _____

Light is not the only form of energy that travels at the speed of light. Radio waves, X-rays, infrared, visible light: all of these have no substance, are wave energy and travel at roughly 300,000 kilometres per second. Together they make up the **electro-magnetic spectrum**, also known as E-M radiation. Use Unit 4.12 to complete the table below.

Shortest wavelengths

Longest wavelengths

Name of E-M 'region'	What are some of its effects on living things (if any)?	What instruments can be used to detect or measure it?
gamma and X-rays		
ultra-violet light		
visible light		
infrared		
microwaves		
radio waves		

17 Emit or reflect?

SP1 Unit 4.12/4.13

Date for completion: / /

Parent sig: _____
Teacher sig: _____

Complete the table below by classifying each of the following as a light **source** (emitter) or a light **reflector**.

the Sun   bathroom mirror   glow-worms   the moon
white paper   lamp   back of a CD   lightning
cat's eyes at night computer screen   stars   road signs

Light sources	Light reflectors

Complete the diagrams below by adding missing parts.
For the flat mirror diagrams, you need to add:
- *the normal (shown with a dotted line)*
- *the other light ray (incident or reflected)*
- *measure the angle of incidence and angle of reflection.*

For the curved mirror diagrams, you need to add:
- *the reflected light rays.*

Use a ruler for all light rays. Use a protractor to get exact angles.

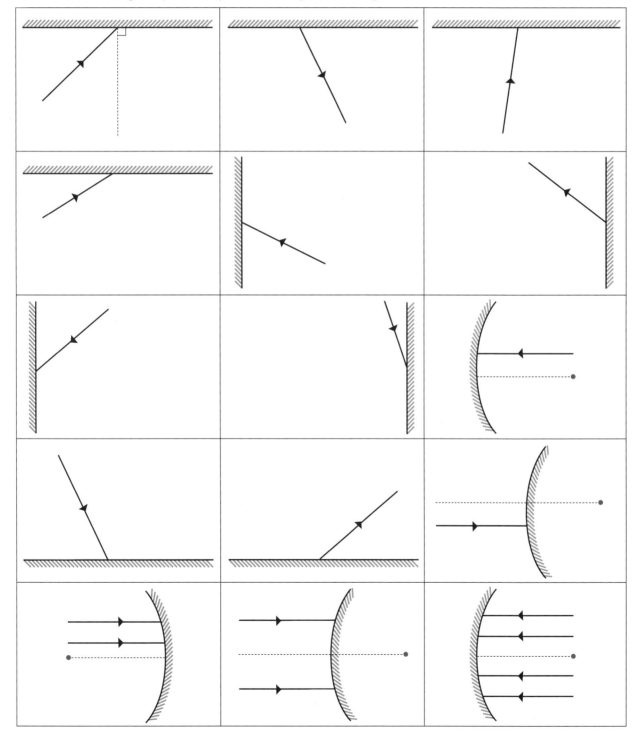

ISBN: 9780170214650

Physical Science

When you look at any image in a flat mirror it is always inverted, which means upside-down (or turned around, depending on how the mirror is held). This is why it's difficult to read writing reflected in a mirror. However, some uppercase letters look the same if they are inverted by a mirror placed on the page just above them. This means that some words look the same as the original when reflected (and upside down).

Examples: C̄H̄Ō̄ĪC̄Ē and B̄ĪK̄Ē, but not RUG.

1 Draw each of the individual letters inverted above the line. You could place a mirror on the line below to check if you are correct:

A B C D E F G H I J K L M N O P Q R S T U V W X Y Z

CHALLENGE

2 Make a list of as many words as possible that look exactly the same when reflected in a mirror placed above. Example: CHOICE. This means you can use only nine uppercase letters. *A list of 20 'mirror image' words is good, 50 is amazing.*

_____
_____
_____
_____
_____
_____
_____
_____
_____
_____
_____

Physical Science

1 Explain the difference between reflection and refraction. Use diagrams if these help you explain.

_____
_____
_____
_____
_____

2   On drawings A, B and C extend the light rays to show what happens when they enter and exit each lens. Draw each ray with a ruler, not freehand. Mark and label the 'focal point' where it applies.

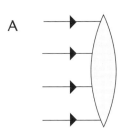

3   Three parallel rays of red light shine through a transparent object and are refracted, as shown below. What is the shape of the object causing this refraction? _____

A   a rectangular glass block
B   a converging (convex) lens
C   a diverging (concave) lens
D   a triangular glass prism

Draw in the approximate outline of the object you have selected.

4   Three parallel light rays shine onto a converging lens. The scale drawing below shows the results. Each square of the grid is 5 cm X 5 cm.

What is focal length of lens A?

_____

Suppose Lens A is replaced with another converging lens, Lens B. This lens has a focal length of 25 cm. Draw the approximate shape of Lens B compared with Lens A on the grid at right.

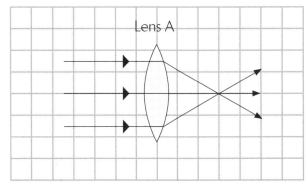

Lens A

21	**Light revision**

Date for completion: / /   Parent sig: _____
Teacher sig: _____

SP1 Unit 4.12–4.15

1   Match each word to the description given: *opaque, refraction, dispersion, reflection, transparent.*

	A substance that allows light rays to pass through.
	Light bouncing off a substance.
	A substance that does not allow light through.
	White light is split into different colours.
	Light passing from air to glass.

ISBN: 9780170214650

2 Name the type of lens shown at right _____

Draw a ray diagram that shows the rays **after** they pass through this lens. Use a ruler. Label the focal point. Give an example of where this type of lens is used.

3 Name the type of mirror shown at right _____

Draw a ray diagram showing the rays **after** reflection from the mirror. Use a ruler. Label the focal point.

22 Colour

SP1 Unit 4.17

Date for completion: / /
Parent sig: _____
Teacher sig: _____

1 What colours are mixed in white light? List what are usually described as 'the seven colours of the spectrum', in order from longest to shortest wavelength.

_____

_____

2 A ray consisting of red, green and blue light shines on a white screen. A coloured filter is placed over the ray. Green light is seen on the screen. Which colour(s) are being transmitted by the filter?

_____.

Which colour(s) are being absorbed by the filter?

_____.

What is the colour of the filter? _____

What will you see on the screen if the filter is replaced with a blue filter? _____

Which colour(s) would you see on the screen if the filter is replaced with a cyan-coloured filter? _____

What colour(s) would you see if you have a blue filter and a red screen? _____

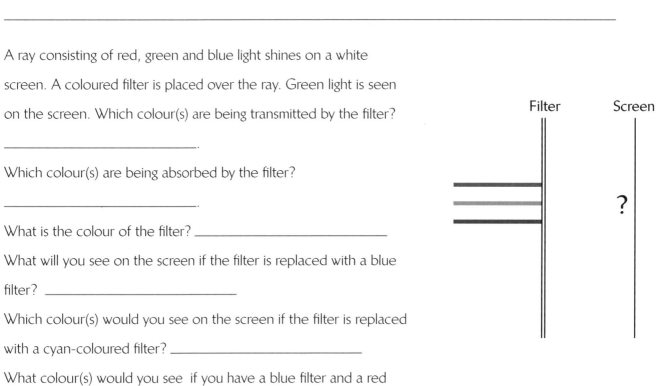

Filter   Screen

?

Physical Science

3 Explain why a banana looks yellow in white light. Use the words *absorbed* and *reflected* in your explanation.

_____

_____

_____

_____

_____

4 What colour would a bunch of green grapes be if you saw them in a red spotlight in a dark room?

_____

5 A beam of white light passes through a filter and then through a prism. The prism refracts the light, and the resulting colours shine on the white screen. Use the information in the diagram at right to decide on the colour of the filter. _____

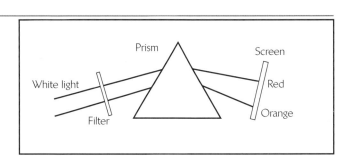

23 Why is the sky blue?

SP1 Unit 4.17

Date for completion: / /   Parent sig: _____
Teacher sig: _____

When you look at the night sky, it's black, with the stars forming points of light on a black background. So why is it that, during the day, the sky doesn't remain black with the Sun acting as another point of light? Why does the daytime sky turn a bright blue and the stars disappear?

The first thing to recognise is that the Sun is extremely bright – much brighter than the moon. The second thing to recognise is that the molecules of nitrogen and oxygen in the atmosphere have an effect on the sunlight that passes through them.

A physical phenomenon called the Rayleigh Effect causes light to scatter when it passes through particles with a diameter about one-tenth that of the wavelength (colour) of light. Sunlight is made up of different colours. The colour blue, because it has a shorter wavelength, is scattered much more than the other colours. Red, with a longer wavelength, is scattered less.

When you look at the sky on a clear day, the Sun is a bright disk. The blueness you see everywhere else is due to particles in the atmosphere scattering blue light in all directions. Because red, yellow and green light aren't scattered nearly as much, you see the sky as blue. *Now answer questions 1 to 5.*

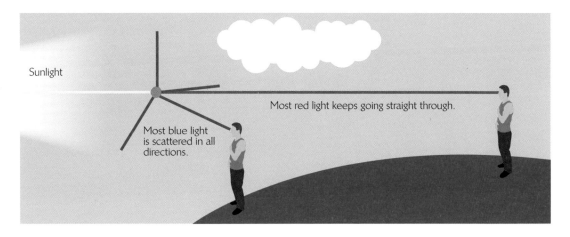

Physical Science

ISBN: 9780170214650

1 Name the two main gases in the atmosphere.

_____

2 Explain what is special about the wavelength of the blue light that makes it scatter more than red does.

_____

_____

3 In photographs taken by astronauts on the moon in full sunlight, the sky looks black. Explain why.

_____

_____

_____

4 Explain what causes the Sun to have a red-orange colour at sunset and sunrise.

_____

_____

_____

5 Sum up the 'Rayleigh effect' in 10 to 20 words

_____

_____

_____

_____

24 Eyes and cameras

Date for completion: / /
Parent sig: _____
Teacher sig: _____

SP1 Unit 4.18

In many ways, the features of a human eye have the same function as features of a camera. In the lists on the next page, connect matching features by drawing straight lines between the two sides. One has been done for you. Several features do not match up: write 'no match' next to any such feature.

Eye		Camera
Cornea and lens		Lens cover
Iris		CCD light sensor, or film
Eyelashes		Glass or plastic lens
Eyelid		Automatic flash
Blood supply		Diaphragm and aperture
Retina		Battery
Ciliary muscle		Memory card
Optic nerve		Automatic focus
Tear gland		Black interior
Black layer behind retina		Zoom adjustment

25 Lasers

Date for completion: / /   Parent sig: _____ Teacher sig: _____

Physical Science

Lasers were first developed in 1960, and new uses are still being found for them. The name is an acronym for 'Light Amplification by the Stimulated Emission of Radiation'. Laser-generators can produce a narrow high-energy beam of light that does not widen with distance, as other light beams do. The waves are always coherent ('in step') and do not weaken over distance. Different types of laser produce specific wavelengths for different purposes. Here are some uses for lasers.

A **Cutting metals.** The fine beam of a laser light can be used for accurate cutting and drilling. Heat from the beam vaporises the metal. Some kinds of laser will slice through steel plate in seconds.

B **Surgery.** Fine laser beams are useful for surgical cutting. They will vaporise tissue without hurting healthy tissue that the beam is not touching. Their heat will sear broken blood vessels and prevent bleeding. They are often used for removing birthmarks, and also for eye surgery.

C **Measuring distance.** Pulses of laser light are sent to purpose-made reflectors on the moon. They reflect back without 'getting lost' in the atmosphere because they do not spread out. By timing how long it takes to go there and back, we can work out accurately how far away the moon is. Discovery: the moon is slowly moving further away.

D **Communication.** Laser beams inside thin fibre optic cables ('light pipes') are used for telephone conversations. These are much less bulky than copper wires, and can carry thousands more conversations than copper is able to.

E **Entertainment.** In some amusement parks, people hunt each other with laser 'guns'. They fire a low-power laser beam that activates a device on the target if the shot is accurate, which tells them they have been 'shot'!

F **Checkouts.** Supermarkets use low-power laser beams to read the barcodes on goods.

ISBN: 9780170214650

1 Choose any three uses of lasers described above that you found the most interesting or useful. Explain briefly **why** each seems especially interesting or useful.

_____

_____

_____

_____

_____

_____

_____

_____

_____

_____

_____

_____

_____

_____

2 Write a short paragraph explaining the differences between laser light and ordinary light.

_____

_____

_____

_____

_____

_____

_____

26 What is sound?

Date for completion: / /

Parent sig: _____

Teacher sig: _____

SP1 Unit 4.19

1 Use the six words given here (plus any others you choose) to write 25 to 50 words explaining the physical basics of sound: _compression, air particles, wavelength, energy, wave._

_____

_____

_____

_____

_____

2 Place these substances in order from the ones that 'transmit' sound, slowest to fastest.

A Steel
B Ice
C Sea water
D Air at sea level
E Air at mountaintop
F Concrete

☐ ☐ ☐ ☐ ☐ ☐

Slowest
because least dense

Fastest
because densest

3 Hannah's brother tells her the reason we hear thunder after we see lightning is that the rain in the sky around the lightning traps the sound for a little while. Is this true? Write Hannah's explanation of the real reasons why we don't see lightning and hear thunder at exactly the same moment.

_____

_____

_____

_____

CHALLENGE

4 Sound travels faster in solids than liquids, and slower in gases. Explain why, in terms of particles.

_____

_____

_____

_____

_____

27 **Human ear**

Date for completion: / /

Parent sig: _____

Teacher sig: _____

SP1 Unit 4.21

Physical Science

1 Label this diagram showing parts of a human ear. This list provides six of the 11 labels you will need: *semicircular canals, eustachian tube, ear canal, cochlea, ear bones, auditory nerve.*

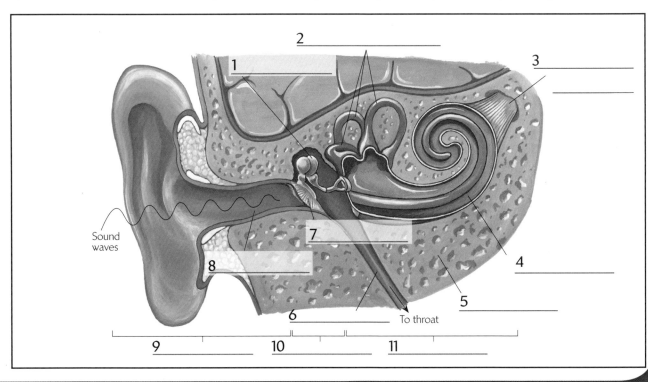

ISBN: 9780170214650

2 From the instant a sound reaches your eardrum, it takes only a few milliseconds before it is registered by your brain. The events A to I below are all part of the process, but are listed in the wrong order. Arrange them in the order in which they occur.

A Sound reaches your eardrum
B Nerve signals taken to brain
C Hair-cell nerve endings are touched
D Stirrup moves
E The hammer and anvil move
F Brain interprets the sound as your friend's voice
G Stirrup vibrates the fluid inside the cochlea
H Eardrum vibrates
I Cochlea fluid causes the tectorial membrane to move

Fact File

Compared to our eyes, human hearing is sensitive to incredibly small energy inputs. A young person with good hearing can detect 0.01 million millionth of a joule of sound energy. (That's 10^{-14} J.) If you part-cover your ears with your hands (or a seashell) when in a quiet place, the 'whispering' you hear is actually the sound of air molecules colliding with your eardrums.

28 Cochlea implants

Date for completion: / /   Parent sig: _____
Teacher sig: _____

A cochlea implant (CI) is a surgically implanted electronic device that provides a sense of sound to a person who is profoundly deaf. The first attempt to develop a clinical CI was in 1957, and since then many improvements have taken place after many trials on animals and people.

The external parts – placed behind the ear – consist of a small microphone that picks up sound from the air; a speech processor that prioritises audible speech and sends the electrical sound signals to a transmitter, which is fitted closely to the surface of the head behind the ear. It is held in place by magnets under the skin. The transmitter sends signals through the skin by electromagnetic induction.

The internal implant is placed under the skin behind the ear. It consists of a receiver and stimulator secured in bone beneath the skin, which converts the signals into electric impulses and sends them through an internal cable. This leads to an array of up to 24 microscopic platinum electrodes carefully placed in the cochlea. The cochlea can then send nerve impulses directly to the brain. The device is surgically implanted under a general anaesthetic, and the operation usually takes from 1½ to 5 hours.

A child can be a candidate for CI only if he/she is profoundly deaf in both ears, has functioning auditory nerves, and a family willing to work toward helping their child with speech and language. The quality of sound heard is not good, since 24 electrodes replace the 16,000 delicate hair cells that are used for normal hearing. While cochlear implants restore some physical ability to hear, this does not mean the brain can automatically process and distinguish speech. Brain pathways for understanding sound develop in childhood.

Physical Science

This means that CIs are useful if implanted in younger children. If a born-deaf adult is given a CI, it will help them hear sounds but they can make little sense of human speech. This is because neural patterns are laid down in the early years of life, which are crucially important to perception.

CI can fail for a number of other reasons, one of which is that during surgery it may be discovered that the auditory nerve is not functional. Some effects of implantation are irreversible; as the implantation process results in damage of the hair cells within the cochlea, this can result in a permanent loss of residual natural hearing ability.

The strongest objections to cochlea implants have come from the deaf community, which consists largely of born-deaf people whose first language is a signed language. For some, cochlear implants are an affront to their culture. They view themselves as a minority threatened by the hearing majority.

1 List four factors that could make a person a suitable candidate for a CI.

_____

_____

_____

_____

2 Write a short description (40 to 80 words) of the external and internal parts of a CI.

_____

_____

_____

_____

_____

_____

_____

_____

_____

_____

_____

ISBN: 9780170214650

3 Explain why a CI is not likely to succeed if given to a 20 year old who has been deaf since birth.

_____

_____

_____

_____

CHALLENGE

4 Write two of your own questions on aspects of CI that are not dealt with in the report.

_____

_____

_____

_____

29 Decibels

Date for completion: / /

Parent sig: _____

Teacher sig: _____

The decibel scale is a measure of the amount of energy in sound waves. The scale is not linear, like metres or kilograms are. The decibel scale (dB for short) is logarithmic, which means it is based on powers of 10. Each increase of 10 in the dB scale represents 10 times as much energy. This means that compared to a 50 dB sound, a 60 dB sound has 10 times as much energy, and 70 dB has 100 times as much energy. But human ears don't measure energy, they experience loudness. To most people, each increase of 10 dB is experienced as about twice as loud. So compared to 50 dB, 60 dB seems twice as loud, and 70 dB seems four times as loud.

1 Complete this table, which compares the energy and loudness from 0 dB (the softest sound that anyone can hear) up to 150 dB (which can break the bones in your ear if you have no hearing protection).

Decibel scale	The amount of energy, comparatively	Perceived loudness, comparatively
0 dB	1	1
10 dB	10	2
20 dB		
30 dB		
40 dB		
50 dB		
60 dB		
70 dB		
80 dB		
90 dB		
100 dB		
120 dB		
140 dB		
150 dB		

165

155

145 — Fireworks Gun shot (close range)

135 — Jet plane at take off (from 30 m away)

125 — Ambulance Jack hammer

— Pain threshold

115 — Leaf blower Rock concert Chainsaw

105 — Loud MP3 player

95 — Lawn mower Hair dryer

85 — Vacuum cleaner

75 — Washing machine Busy traffic

65

55 — Normal conversation

45 — Rainfall

35 — Quiet room

25 — Whisper

— Cat purring

15

5

0 — Softest sound anyone can hear

2 Use the table and diagram on page 126 to answer these questions.

  A Compared to a washing machine, how much louder is a lawnmower? _____

  B Compared to a quiet room, how much louder is a chainsaw? _____

  C Compared to a vacuum cleaner close-up, how many times more energy is in the sound from a jet plane at takeoff, heard from 30 m away? _____

CHALLENGE

3 Being close to a noise of more than 91 dB for two hours a day over a long time is medically known to cause hearing loss. (The hair cells in the cochlea become damaged, and do not regrow.) Nathan listens to loud MP3 music for an average of three hours a day, and says his hearing is fine. Suggest at least two possible reasons for his opinions that his music is doing no harm.

_____

_____

_____

_____

---

30 Music

| Date for completion: | / / | Parent sig: _____ |
| | | Teacher sig: _____ |

SP1 Unit 4.22

1 Fill in the following table to indicate the feature in each instrument that is mainly responsible for creating sound. (Write yes or no in each box.)

	Strings vibrate, compressing the air	Air inside it vibrates
Piano		
Drum		
Recorder		
Violin		
Pipe organ		

2 Make a drawing of a single guitar string close-up, showing it vibrating and creating pressure waves in the air. Also draw air molecules – greatly exaggerated in size. Add words to explain exactly how the string causes the air to vibrate.

Physical Science

ISBN: 9780170214650

_____
_____
_____
_____
_____
_____
_____

_____
_____
_____
_____

3 Complete the sentences using four of these words and numbers: *octave, hertz, loudness, amplitude, 257, 512, 2560, 259, 1024, frequency.*

Musical pitch is known in scientific terms as _____ . Humans can hear up to a

limit of about 18,000 waves per second, a unit also known as _____ . One

musical octave represents a doubling of wave frequency. Middle C has a frequency of 256 Hz, so a

note of C one octave above middle C has a frequency of _____ Hz, and C

two octaves above middle C as a frequency of _____ Hz.

ISBN: 9780170214650

Earth Science

1 Geosphere

Date for completion: / /

Parent sig: _____

Teacher sig: _____

SP1 Unit 5.1–5.2

> 'Earth is a ball of iron covered with a thin layer of rock, and even thinner layers of water and air.' (Wikipedia.com).

1 Label regions A to F in the diagram using terms from this list: *hydrosphere; core; inner mantle; atmosphere; crust; outer mantle.*

A _____

B _____

C _____

D _____

E _____

F _____

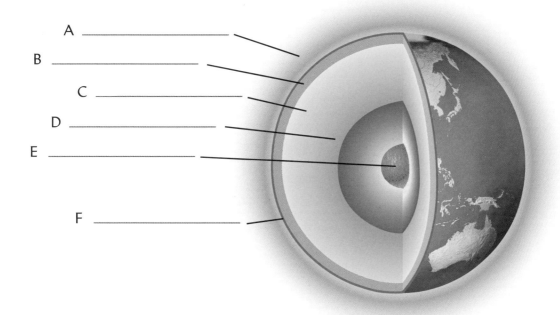

Note: the thicknesses of the crust and atmosphere have been exaggerated in this diagram.

2 Use your textbook to list four kinds of evidence that tell us Earth has only a thin crust of solid rock, and that most of its interior is hot melted rock and iron.

- _____

- _____

- _____

- _____

Earth Science

ISBN: 9780170214650

1 Use the following descriptions to place and label each of the 14 terms in **bold** in its correct place in the diagram.

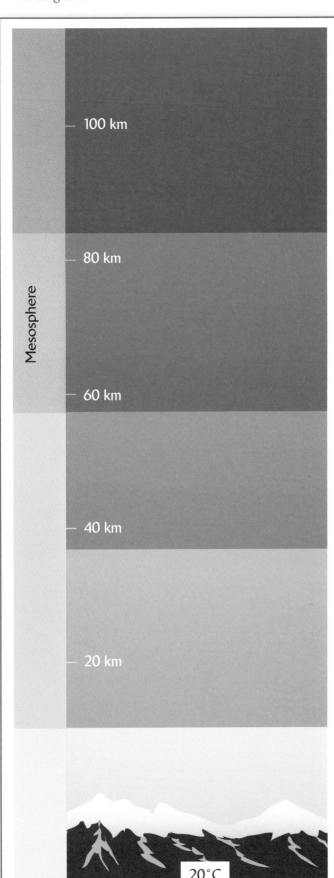

Most **clouds** are below 12 km altitude, in the layer of denser air known as the **troposphere**. The highest **mountains** reach to 8 km, where temperatures are around **-40°C** Jet **aircraft** generally fly at 10 km altitude, while artificial **satellites** are mostly above 100 km. The **stratosphere** is the layer above the troposphere, reaching out to 60 km. Most of the **ozone layer** is in the middle of this layer, at about 20–50 km altitude. Above the stratosphere and up to 80 km is a layer called the **mesosphere** (meso means middle), where temperatures are as low as **-100 °C**. Above this is the **thermosphere** (thermo means heat), where temperatures rise again to as warm as **-10 °C** because of the heating effect of sunlight. Air in this layer becomes thinner and thinner with altitude, up to 400 km and more. There is no clear limit between the atmosphere and **space**. Most incoming **meteoroids** vaporise in this layer.

ISBN: 9780170214650

Earth Science

2 Two possible causes have been suggested for global climate change: (A) An increased heat output from the Sun, which is causing the entire atmosphere to warm up; (B) Human activity, especially the production of greenhouse gases. Careful measurements show that the boundary between troposphere and the stratosphere has moved upwards in recent years and the troposphere is getting warmer, and the stratosphere is not. Do these facts back up suggestion A or suggestion B? Give reasons for your answer.

_____

_____

_____

_____

_____

3 Hydro, geo, astro, bio

Date for completion: / /

Parent sig: _____
Teacher sig: _____

SP1 Unit 5.1

We can look at Earth as consisting of different 'spheres', usually named *atmosphere*, *hydrosphere* and *geosphere*. Sometimes the word *biosphere* is also used to include all living things.

Answer each description A to M by writing one of the four words given in italics above.

A All the gases that surround Earth _____

B The solid parts of the Earth: rocks, crust, mantle, core _____

C All water in the oceans, rivers, lakes, air _____

D A layer about 400 km thick _____

E Mostly in water, or in the top few metres of soil, or just above _____

F Has a diameter of about 13,000 km _____

G Contains the ozone layer _____

H All weather happens here _____

I Covers most of the earth, and influences weather worldwide _____

J Plants make up a major part of it _____

K Volcanoes emerge from it _____

L Consists mostly of iron _____

M Consists mainly of nitrogen _____

Earth Science

ISBN: 9780170214650

Atmosphere, hydrosphere, geosphere and biosphere affect each other in complex ways.

A   Identify two ways a volcanic eruption affects the atmosphere.

_____

B   Identify two ways a volcanic eruption can affect the biosphere.

_____

C   Identify two ways that living things affect the atmosphere.

_____

D   Identify two ways that the atmosphere affects living things.

_____

E   Identify one way that living things affect the geosphere.

_____

_____

F   Identify one way that the geosphere affects living things.

_____

_____

G   Identify one way that the hydrosphere affects living things.

_____

_____

H   Identify one way that the hydrosphere affects the geosphere.

_____

_____

Earth Science

5 Scaled down

Date for completion: / / Parent sig: _____ Teacher sig: _____

On a scale where 1 mm represents 50 km, our 13,000 km diameter Earth would be slightly bigger than a basketball. Use this 50 km→ 1mm scale to calculate, then complete the table.

Feature	Actual measurement	Scale measurement (mm)
Diameter of Earth	13,000 km	
Atmosphere extends up to	150 km	
Height of ozone layer above ground	20 to 50 km	
Height of Mount Everest	8.8 km	
Average depth of the sea	3 km	
Deepest part of the sea	10 km	
Deepest mine	4 km	
Average thickness of Earth's crust	50 km	1 mm
Distance from surface to Earth's centre		
Diameter of the Earth's core	2,000 km	

6 Humidity

Date for completion: / / Parent sig: _____ Teacher sig: _____

The table on the next page gives measurements of **absolute humidity** in g/m³ (grams of water vapour per cubic metre of air). 100% **relative humidity** means that air cannot hold any more water vapour at that temperature. 80% relative humidity means that the air is holding 8/10 of the maximum. (Humidity also depends on air pressure, but the table does not show this.)

Earth Science

ISBN: 9780170214650

Air temp. (°C)	Relative humidity									
	10%	20%	30%	40%	50%	60%	70%	80%	90%	100%
+40	5.1	10.2	15.3	20.5	25.6	30.7	35.8	40.9	46.0	51.1
+35	4.0	79	11.9	15.8	19.8	23.8	27.7	31.7	35.6	39.6
+30	3.0	6.1	9.1	12.1	15.2	18.2	21.3	24.3	27.3	30.4
+25	2.3	4.6	6.9	9.2	11.5	13.8	16.1	18.4	20.7	23.0
+20	1.7	3.5	5.2	6.9	8.7	10.4	12.1	13.8	15.6	17.3
+15	1.3	2.6	3.9	5.1	6.4	7.7	9.0	10.3	11.5	12.8
+10	0.9	1.9	2.8	3.8	4.7	5.6	6.6	7.5	8.5	9.4
+5	0.7	1.4	2.0	2.7	3.4	4.1	4.8	5.4	6.4	6.8
0	0.5	1.0	1.5	1.9	2.4	2.9	3.4	3.9	4.4	4.8
-5	0.3	0.7	1.0	1.4	1.7	2.1	2.4	2.7	3.1	3.4
-10	0.2	0.5	0.7	0.9	1.2	1.4	1.6	1.9	2.1	2.3

Use the table to answer the following:

1 At 20 °C and 100% humidity, how many g/m³ water vapour does air hold? _____

2 At 40 °C and 100% humidity, how many g/m³ water vapour does air hold? _____

3 At 20 °C and 50% humidity, how many g/m³ water vapour does air hold? _____

4 At –5 °C and 50% humidity, how many g/m³ water vapour does air hold? _____

5 At 20 °C and 80% humidity, how many g/m³ water vapour does air hold? _____

6 Generally, which kind of air can hold more water: warm or cold? _____

7 Calculate 80% of your answer to question 1.

_____

How does this calculated answer compare to your answer for question 5?

_____

8 A warm wind has air at 25°C and 100% humidity. How many g/m³ water vapour does this air hold?

_____. A mountain deflects this same 100% humid air up to where the temperature

is 10°C. How many g/m³ can the air now hold in vapour form? _____ Calculate the

amount of water per m³ that will condense as cloud or rain at this new temperature of 10°C.

_____

Earth Science

ISBN: 9780170214650

Wind direction

Date for completion: / /

Parent sig: _____
Teacher sig: _____

SP1 Unit 5.6

Write in the missing words below with ones chosen from this list (some words will be used more than once): *rise, isobars, gentle, pressure, low, Coriolis, strong, dense, spiral, expands, high.*

When air is warmed by the Sun it _____ (1), which makes it become less

_____.(2) As a result, warm air tends to _____ (3) .

This causes a local area of_____ (4) pressure air in the surroundings. Cooler

air moves in from _____ (5) pressure areas to take the place of rising air. Air

_____ (6) can be measured in a variety of ways such as bars and hectopascals.

_____ (7) is the word for lines that connect points which have the same air

pressure at ground level. These lines generally form curves and circles centred on regions of high

_____ (8) and low _____ (9). Winds do not blow straight

from high to low, but blow almost parallel to the _____ (10) lines. Because of

the _____ (11) effect, caused by Earth's rotation, wind systems tend to form

_____ (12) patterns, which we can see in the case of tropical cyclones. When

isobar lines are close together, _____(13) winds are certain; when these lines are

further apart, _____ (14) winds are likely.

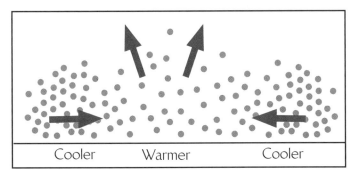

Cooler   Warmer   Cooler

Warm air expands. This causes pressure to become less.
Cold air is dense. This causes pressure to become greater.

Earth Science

ISBN: 9780170214650

8 Rainfall patterns

Date for completion: / /

Parent sig: _____

Teacher sig: _____

Use the table below to answer the following questions.

MEAN MONTHLY RAINFALL (mm)

Location	Jan	Feb	Mar	Apr	May	Jun	Jul	Aug	Sep	Oct	Nov	Dec	Year
Kaitaia	82	79	78	95	119	149	166	152	133	93	94	97	1334
Whangarei	90	112	142	129	120	179	151	146	130	116	80	92	1490
Auckland	75	65	94	105	103	139	146	121	116	91	93	91	1240
Tauranga	74	78	128	105	91	128	122	115	104	94	85	87	1198
Hamilton	85	71	87	95	102	119	126	117	102	96	93	95	1190
Rotorua	99	101	115	112	104	134	130	148	119	122	102	115	1401
Gisborne	54	78	99	103	97	125	119	93	101	63	65	67	1050
Taupo	85	77	83	74	87	99	105	109	90	102	85	108	1102
New Plymouth	97	95	117	131	124	145	143	127	110	124	108	103	1432
Napier	48	62	85	75	62	81	92	67	65	55	57	56	803
Wanganui	62	65	68	71	81	82	88	70	72	81	74	70	882
Palmerston North	65	62	74	76	94	87	94	82	83	90	78	83	966
Wellington	72	62	92	100	117	147	136	123	100	115	99	86	1249
Nelson	72	57	78	86	77	85	86	90	73	92	82	75	970
Blenheim	47	27	54	64	58	56	71	70	44	70	43	54	655
Kaikoura	47	59	92	81	71	75	80	78	70	74	60	54	844
Hokitika	250	172	217	249	245	233	232	224	250	286	240	278	2875
Christchurch	42	39	54	54	56	66	79	69	47	53	44	49	648
Aoraki/Mt Cook	411	255	422	362	365	287	278	298	310	452	390	461	4293
Timaru	46	38	52	66	42	41	43	45	35	55	48	53	573
Milford Sound	717	499	640	585	641	440	418	427	523	688	522	648	6749
Queenstown	78	58	80	75	89	82	65	73	69	95	72	77	913
Alexandra	29	22	40	34	35	26	23	24	27	41	26	43	360
Dunedin	72	63	70	60	72	74	69	65	53	71	63	82	812
Invercargill	114	79	94	100	114	99	88	71	80	95	81	100	1112

1 Which of the listed places in New Zealand has the lowest average annual rainfall?

_____ How much, in millimetres? _____

2 Which of the listed places in New Zealand has the highest average annual rainfall?

_____ How much, in millimetres? _____

And in metres? _____

3 Milford Sound and Queenstown are only 60 km apart, and both are surrounded by mountains. Explain why one of these places gets seven times as much total rainfall as the other. Use a drawing as part of your explanation.

_____

_____

_____

_____

_____

_____

_____

_____

_____

_____

_____

4 For any weather station or town of your choice, draw a column graph showing monthly variation in rainfall. Remember the TADPL (Title, Axes, Divisions, Plot, Lines) guidelines (see page 15).

ISBN: 9780170214650

Date for completion: / /

Parent sig: _____

Teacher sig: _____

On the blank map below, mark all 25 places named in the table in Earth Science Unit 8. If necessary, use an atlas to help you.

Earth Science

Date for completion: / /

Parent sig: _____

Teacher sig: _____

SP1 Unit 5.8

KEY

Isobars

Cold front

Warm front

Wind arrow

1 What is the highest air pressure shown on the above map?_____

The lowest?_____

On the map, draw wind direction arrows for: North Cape; Coromandel; Napier; Invercargill; and the central Tasman Sea. To each arrow, add one tail feather for light winds, two for moderate or four for very strong winds.

2

Region	Evidence or reason, based on the weather map	Prediction (forecast) from this evidence
Coromandel	Deep low nearby. Isobars are very close spaced	Very strong winds. Direction: _____
Napier		Wind strength:_____ Direction:_____
		(Chances of rain?) _____
Wellington region		(Wind strength? Rain?) _____
West Coast of South Island		(Wind strength? Rain?) _____
Christchurch		Cooler rainy weather in a few hours
Tasmania		

Earth Science

ISBN: 9780170214650

The table below shows mean (average) monthly temperatures. Use it to answer the following questions.

MEAN MONTHLY AIR TEMPERATURE (°C)

Location	Jan	Feb	Mar	Apr	May	Jun	Jul	Aug	Sep	Oct	Nov	Dec
Kaitaia	19.7	20.0	18.6	16.9	14.8	12.7	12.2	12.1	13.1	14.5	15.8	17.9
Whangarei	19.9	20.0	19.0	16.5	14.0	12.2	11.2	11.7	12.9	14.3	16.4	18.2
Auckland	19.3	19.8	18.5	16.2	13.7	11.6	10.8	11.3	12.6	14.1	15.8	17.8
Tauranga	19.2	19.2	17.9	15.4	12.5	10.4	9.7	10.5	12.1	13.6	15.6	17.5
Hamilton	18.3	18.7	17.1	14.5	11.6	9.4	8.7	9.8	11.4	13.1	15.0	16.8
Rotorua	17.8	17.9	16.4	13.5	10.5	8.4	7.6	8.7	10.4	12.3	14.3	16.1
Gisborne	19.2	18.9	17.4	14.8	12.0	10.0	9.3	10.2	11.8	13.8	15.9	17.8
Taupo	17.4	17.5	15.6	12.5	9.6	7.6	6.7	7.6	9.4	11.5	13.6	15.7
New Plymouth	17.7	17.9	16.9	14.7	12.1	10.3	9.4	10.1	11.5	12.7	14.4	16.1
Napier	19.5	19.3	17.7	15.1	12.1	9.7	9.3	10.0	12.0	14.3	16.1	18.2
Wanganui	18.2	18.3	17.1	14.7	12.3	10.2	9.4	10.0	11.7	13.3	14.9	16.7
Palmerston North	17.9	18.2	16.6	14.1	11.3	9.2	8.6	9.3	11.0	12.7	14.3	16.3
Masterton	17.8	17.6	16.1	13.3	10.4	8.2	7.5	8.4	10.3	12.1	14.0	16.1
Wellington	16.9	17.1	15.8	13.8	11.5	9.5	8.8	9.2	10.6	12.0	13.4	15.3
Nelson	17.7	17.7	16.1	13.1	10.1	7.7	7.0	8.1	10.2	12.3	14.2	16.2
Blenheim	18.2	17.9	16.7	13.7	10.4	7.9	7.4	8.5	10.8	12.8	14.9	16.7
Westport	16.2	16.5	15.7	13.5	11.0	9.2	8.6	9.3	10.7	11.8	13.4	15.1
Kaikoura	16.7	16.4	15.3	13.3	10.9	8.7	8.0	8.5	10.1	11.7	13.4	15.4
Hokitika	15.6	16.0	14.7	12.5	10.0	8.0	7.5	8.2	9.8	11.3	12.7	14.3
Christchurch	17.4	17.1	15.5	12.8	9.6	6.9	6.6	7.7	10.0	12.3	14.0	16.0
Aoraki/Mt Cook	14.6	14.8	12.5	9.4	6.1	3.0	2.2	3.7	6.6	8.9	10.8	12.8
Lake Tekapo	15.2	15.1	12.7	9.5	6.0	2.5	1.7	3.4	6.6	9.1	11.1	13.3
Timaru	16.2	16.0	14.7	11.9	8.6	6.0	5.7	6.8	9.4	11.2	13.4	15.1
Milford Sound	14.7	15.0	13.5	11.2	8.3	5.7	5.2	6.7	8.5	10.1	11.7	13.4
Queenstown	16.7	16.6	14.4	11.1	7.6	4.6	4.1	5.8	8.6	10.9	13.0	15.1
Alexandra	17.1	17.1	15.1	11.3	7.0	2.9	3.0	5.5	8.9	11.5	13.9	16.0
Manapouri	14.5	14.3	11.8	9.2	7.2	4.2	3.9	4.7	7.1	9.4	10.3	12.8
Dunedin	15.2	15.1	13.7	11.9	9.2	7.0	6.5	7.5	9.3	10.9	12.4	13.9
Invercargill	14.0	13.9	12.5	10.4	8.0	5.6	5.2	6.4	8.3	10.0	11.3	13.0
Chatham Islands	14.7	15.1	14.2	12.5	10.2	8.7	8.0	8.5	9.3	10.5	11.8	13.7

Earth Science

ISBN: 9780170214650

1   Which place has the highest January average?   _____ ; _____ °C

2   Which place has the coldest July average?   _____ ; _____ °C

3   Which place has the smallest difference between January and July average temperatures?

    _____.   What is this difference in °C ? _____

4   Use the table to draw two line graphs showing monthly averages for Dunedin and Alexandra.

5   What is the January-July difference for Dunedin?   _____°C.

    What is the January-July difference for Alexandra? _____ °C

6   Explain in detail why one of these places has a big January-July temperature difference, and the other
    has a small difference.

    _____

    _____

    _____

    _____

7   Everyone knows that summer is warmer than in winter. Identify two underlying causes of this
    temperature difference. (Note: we are **not** nearer the Sun in summer!)

    _____

    _____

    _____

    _____

Earth Science

ISBN: 9780170214650

Earth Science

El Niño is a natural feature of the global climate system. Originally it was the name given to the periodic development of unusually warm sea waters along the South American coast, but now it is generally used to describe the whole El Niño-Southern Oscillation (ENSO) phenomenon, the major global climate fluctuation that occurs at the time of an ocean warming event. El Niño and La Niña refer to opposite extremes of the ENSO cycle, when major changes in oceanic circulation occur. In South America, these events tend to begin around Christmas, hence the name El Niño, meaning Christ child.

During El Niño events there is a rise in sea surface temperature in the eastern equatorial Pacific, and a reduction of up-welling off South America. Heavy rainfall and flooding occur over Peru, and drought over Indonesia and Australia. The supplies of nutrient-rich water from the South American coast are cut off due to the reduced up-welling, and the numbers of fish being caught are greatly reduced. In the tropical South Pacific the pattern of occurrence of tropical cyclones shifts eastward, so there are more cyclones than normal in areas such as the Cook Islands and French Polynesia.

During La Niña events the trade winds strengthen, with even colder sea surface temperatures in the eastern Pacific, and more tropical cyclones in the western Pacific. These events generally happen every 10 to 12 years.

During El Niño, New Zealand tends to experience stronger or more frequent winds from the west in summer, typically leading to drought in east coast areas and more rain in the west. In winter, the winds tend to come more from the south, bringing colder conditions to both the land and the surrounding ocean. In spring and autumn southwesterly winds are more common.

La Niña events have different impacts on New Zealand's climate. More northeasterly winds are characteristic, which tend to bring moist, rainy conditions to the northeast of the North Island, and reduced rainfall to the south and southwest of the South Island. Therefore some areas, such as central Otago and South Canterbury, can experience drought in both El Niño and La Niña. Warmer than normal temperatures typically occur over much of the country during La Niña, although there are regional exceptions.

Although ENSO events have an important influence on New Zealand's climate, it accounts for less than 25% of the year-to-year variance in seasonal rainfall and temperature at most New Zealand measurement sites. East coast droughts may be common during El Niño events, but they can also happen in non La Niña years.

ISBN: 9780170214650

1 What does 'ENSO' stand for? _____

2 List three effects that El Niño events have on the South Pacific, Australia, and on South America.

_____

_____

_____

3 List four effects that El Niño events tend to have on New Zealand.

_____

_____

_____

_____

4 List two effects that La Niña events tend to have on New Zealand.

_____

_____

13 Greenhouse effect

Date for completion: / /

Parent sig: _____

Teacher sig: _____

SP1 Unit 5.3/5.10

In the sentences below, write the following words and figures (in places the first letters will help you): *methane, storms, radiate, weather, fossil, 30°C, absorbed, 280 ppm, icecaps, evidence, twelve, droughts, crop, reflected, CO_2, longer, tropical, 15°C, floods, 380 ppm, carbon.*

About one third of the sunlight reaching Earth is r_____ (1) off clouds before it even reaches the Earth is surface. Much of the remaining light that does reach the surface is a_____ (2), warming both land and sea. At night, both land and sea ra_____ (3) infrared light which has a l_____ (4) wavelength than the incoming light. Most of this infrared radiation is sent out into space again at the speed of light, but part of it is absorbed by clouds and some gases in the atmosphere, which traps heat and radiates it back to Earth again. This is called the greenhouse effect. The two main greenhouse gases are _____ (5), and m_____ (6) This warming effect is natural, and has for thousands of years kept the average Earth temperature at about _____ (7) We calculate that if the greenhouse effect didn't exist, Earth's temperature would be about _____ (8) colder.

Continued over.

Earth Science

ISBN: 9780170214650

Problem is, the amount of greenhouse gases, especially _____ (9), but also

_____ (10), have been rising. At the start of the industrial age the concentration

of _____ (11) in the air was _____ (12) It is now almost

_____ (13), mostly as a result of burning _____ (14) fuels.

By 2010, this 'new' human activity was putting about _____ (15) million tonnes of

carbon dioxide into the air every day. The natural working of the_____ (16) cycle

has been altered. Science cannot prove the future, but most e_____ (17) points

to a real risk that an increased greenhouse effect will cause Earth's average temperature to rise by

anywhere from 2 °C to 6 °C, causing w_____ (18) patterns to change. The

main flow-on effects are likely to be more _____ (19) more

_____ (20), and more _____ (21), plus spreading

t_____ (22) pests, and increased risk of_____ (23) failures.

Possible rises in sea level from partly-melted _____ (24) would bring an extra set

of problems.

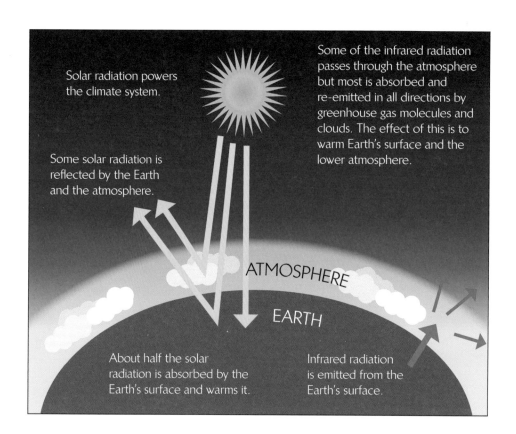

Solar radiation powers the climate system.

Some of the infrared radiation passes through the atmosphere but most is absorbed and re-emitted in all directions by greenhouse gas molecules and clouds. The effect of this is to warm Earth's surface and the lower atmosphere.

Some solar radiation is reflected by the Earth and the atmosphere.

ATMOSPHERE

EARTH

About half the solar radiation is absorbed by the Earth's surface and warms it.

Infrared radiation is emitted from the Earth's surface.

Astronomy

1   Day and night

Date for completion: /   /

Parent sig: _____

Teacher sig: _____

SP1 Unit 5.12

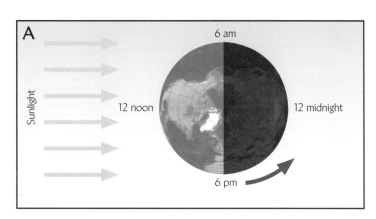

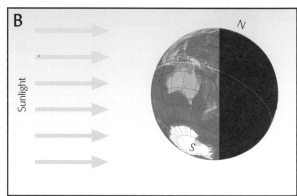

1   How many days does it take Earth to complete one orbit of the Sun?

_____

2   Explain what causes day and night.

_____

3   Does Earth rotate from west to east, or east to west?

_____

4   In drawing B, is New Zealand close to sunrise, or to sunset?

_____

5   Explain what is technically wrong with the description 'sunset'.

_____

_____

_____

CHALLENGE

6   Explain why every fourth year is a 'leap' year of 366 days (29th February).

_____

_____

_____

_____

_____

Astronomy

ISBN: 978-017-409384-1

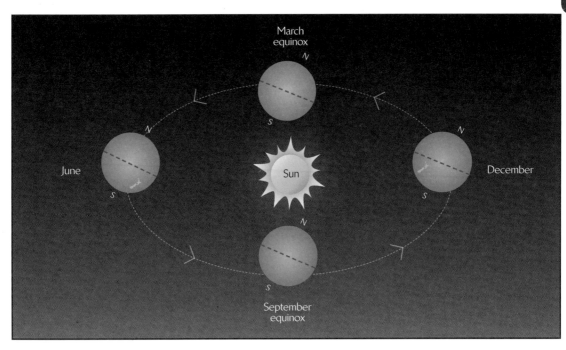

The diagram shows Earth in four stages in its one-year orbit around the Sun.

1 In which of the months shown in the drawing is the Sun highest in the sky in the **southern** hemisphere?

_____

2 In which month is the Sun lowest in the sky in the **northern** hemisphere?

_____

3 In which month is there **no** sunlight at the South Pole?

_____

4 In its orbit around the Sun, Earth travels 107,000 km per hour. Calculate how far it travels in 24 hours; then how far it travels in one year.

_____

_____

_____

5 The equinox dates are 21st March and 21st September. Explain what is meant by 'equinox'.

_____

_____

6 In which of the months shown in the diagram is it spring in New Zealand?

_____

7 On 21st March, how many hours long is the night in Australia _____; and in

Russia _____?

8 Which of the following five time-measures (A to E) are **natural**, and which are completely **human-made**?

A Year _____, because in nature ..._____

_____

ISBN: 9780170214650

Astronomy

B day _____, because in nature ...
_____

C week _____, because in nature ...
_____

D hour _____

E minute _____

Tides

Date for completion: / /   Parent sig: _____ Teacher sig: _____

SP1 Unit 5.12

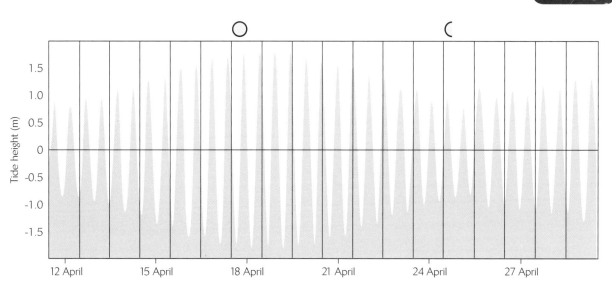

The graph above shows tide height predictions for Aotea Harbour, for 28 days starting on 12 April. Full moon is on 18 April. The first quarter (half-shaded moon) occurs one week later.

1 Identify at least two practical uses of predicting tide heights.

_____

_____

_____

2 Explain what is meant by a 'full moon'.

_____

_____

3 Over the two weeks, on what date was the biggest tide range? (difference between highest and lowest)

_____. About how many metres? _____.

What do we call this kind of tide?_____

4 Over the two weeks, on what date was the smallest tide range (difference between highest and lowest)

_____. About how many metres?_____

What do we call this kind of tide?_____

Astronomy

ISBN: 9780170214650

5 What force from the moon affects the tides?

_____

6 Explain why tides are biggest about the time of a new moon.

_____

_____

_____

_____

CHALLENGE

7 The graph covers 18 days. How many hours is this? _____

Count how many high tides occurred in this time. _____

Calculate the average time between one high tide and next.
Calculation:

_____

Answer: _____ hours

---

4 Moon time

Date for completion: / /

Parent sig: _____
Teacher sig: _____

SP1 Unit 5.12

This diagram shows the cycle of our moon going around Earth. Use it to answer questions 1 to 6.

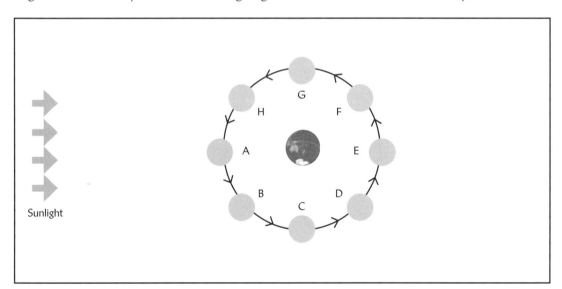

Sunlight

1 How many days does it take the moon to orbit Earth once? _____

2 Which position of the moon could possibly cause a **solar** eclipse? _____

A solar eclipse is when the _____ is briefly hidden by

the _____ moving in front of it.

3 Which position of the moon could possibly cause a **lunar** eclipse? _____

A lunar eclipse is when the shadow of the _____ moves across the

_____.

Astronomy

4 Which moon positions will cause spring tides on Earth? _____

5 Sketch what the moon outline will look like in each of positions A to H, as seen from our position on Earth. Use dark shading to show sunless (hardly-visible) parts of the moon.

A	B	C	D
E	F	G	H

6 Give the letters that correspond to each of these descriptions of moon shapes and positions.

Description	Letter
Full moon	
New moon (invisible)	
First quarter	
Waning (shrinking) crescent	

7 Add to the diagram below, in order to show why the moon looks back-lit (crescent-shaped) to a person standing at X. Extend the ruled lines showing rays of sunlight. Add dark shading to show the exact areas of night-shadow on both Earth and the moon. Label your drawing and explain why the moon looks cresent-shaped.

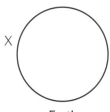

X

Earth

Sunlight

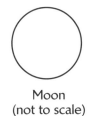

Moon
(not to scale)

Astronomy

ISBN: 9780170214650

Comets and meteoroids are objects that orbit the Sun. Both are a lot smaller than even the smallest planets and there are many thousands of them. Comets are made up of rock and ice, and travel on elongated orbits that can extend far beyond the outermost planets. They tend to return past the Sun at regular intervals many years apart. Bigger comets may stay visible for several weeks, with long tails caused by the Sun's heat evaporating gases from their surfaces. Most meteoroids are fragments of rock and dust. Objects bigger than a few metres diameter qualify as asteroids. If a meteoroid enters Earth's atmosphere, friction causes it to burn out. If this happens at night you may see a streak of light known as a meteor. If a meteoroid reaches the ground without burning up it becomes a meteorite. Asteroids range from 50 metres across, up to the biggest one, Ceres, which has a diameter of over 900 km, almost big enough to qualify as a planet.

Use the above description to answer the following.

1 Identify two differences that would enable you to recognise the difference between a meteoroid and a comet when you actually see them in the night sky.

_____

_____

_____

2 Describe the difference between a meteoroid and a meteor.

_____

_____

3 Under what conditions does a meteoroid become a meteorite?

_____

_____

4 Identify the difference between a meteoroid and an asteroid.

_____

_____

5 Explain what causes a comet's tail.

_____

_____

6 What is unusual about the path of comets, compared to planets?

_____

_____

7 What is the difference between an asteroid and a planet?

_____

_____

8 On this diagram (not to scale), add arrows and names to identify eight planets and one comet.

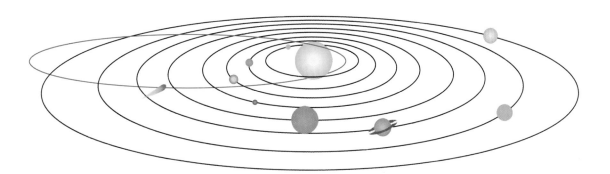

ISBN: 9780170214650

Solar system crossword

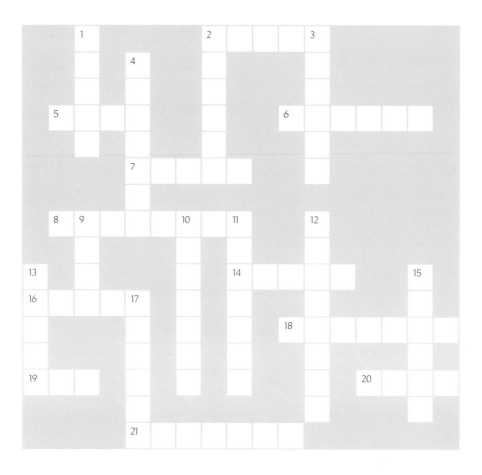

ACROSS

2 The biggest asteroid
5 Tides with small up-down movements
6 First mission to Mars
7 One of the moons of Saturn
8 Makes up over 90% of the universe
14 Demoted to the status of dwarf planet
16 Cool, wet and blue
18 Pulls everything together
19 Centre of the solar system
20 Fourth rock from the Sun, with two moons
21 The smallest planet

DOWN

1 Caused by the pull of the moon and Sun
2 Glowing outer layer of the Sun
3 Also known as king tides
4 Biggest planet of all, with 63 moons
9 How long it takes a planet to complete one circuit
10 A moon that is not quite 'full'
11 Planet named after a Roman god of the sea
12 Space rock from 50 m to 900 km across
13 Planet with an atmosphere of CO_2, and hot enough to melt lead
15 Has rings around it
17 This element was first discovered in the Sun

Write each of these words next to its matching description in the list below: *gravity, orbit, planet, natural satellite, poles, atmosphere, corona, circumference, Sun, equator, light source, axis, rotating*. Each word is used once.

1   Large spherical mass orbiting a star : _____

2   Layers of gases around a planet : _____

3   Imaginary line around Earth midway between the poles : _____

4   Able to generate its own light : _____

5   Imaginary line running through Earth from pole to pole : _____

6   Imaginary points at the top and bottom of a planet : _____

7   Spinning around an axis : _____

8   An elliptical trip around an object : _____

9   Distance around a circle : _____

10 Force of attraction between objects over a distance : _____

11 A moon orbiting a planet : _____

12 Bigger mass than the rest of the solar system : _____

13 Outer layer of the Sun : _____

Jade drew a picture of our solar system by reducing its actual size by a factor of 150 million km. (i.e. the distance from Earth to the Sun). She calculated the distance of each planet from the Sun and recorded it on the table below — but forgot to write the planet names.

Astronomy

ISBN: 9780170214650

Use this table to answer questions 1, 2 and 3. (AU is short for 'astronomical unit'.)

Planet	Distance from the Sun (AU)
A	0.39
B	0.72
C	1.00
D	1.52
E	5.20
F	9.59
G	19.19
H	30.06

1 What could be the names of the planets listed as:

A _____

C _____

G _____

2 Calculate the distance of planet G from the Sun in kilometres (to three significant figures).

_____

3 Which planet is about 780 million kilometres from the Sun? Letter: _____

Now use the data in the next table to answer questions 4 to 7.

Planet	Mercury	Venus	Earth	Mars	Jupiter	Saturn	Uranus	Neptune
Average distance from Sun (AU)	0.39	0.72	1.0	1.5	5.2	9.5	19.2	30.1
Orbital period (Earth years)	0.24	0.62	1.0	1.9	19	29.5	84	165
Surface gravity (Earth = 1.0)	0.38	0.91	1.00	0.38	2.36	1.22	0.89	1.14
Average surface temperature (°C)	350	400	25	-25	-140	-170	-210	-225
Rotational period (Earth days)	59	243	1.0	1.03	9.92	10.7	17.2	16.1

4 Which planet is closer to Earth: Venus or Mars ? _____

5 Describe the relationship between the size of the planets and their surface gravity.

_____

_____

ISBN: 9780170214650

6 What information in the data table shows that the rotation of Venus on its axis is slower than other planets in our solar system?

_____

_____

_____

7 For the six innermost planets, draw a line graph showing orbital period of each planet on the Y-axis, and distance from the Sun for each planet on the X-axis. What relationship does the graph show?

_____

_____

_____

_____

Astronomy

ISBN: 9780170214650

Glossary

absolute (kelvin) temperature scale (K) the SI temperature scale. On this scale, zero is absolute zero and the scale divisions are the same size as those of the Celsius scale

acoustics the science of sound

adapted genetically fitted for a particular environment and way of life

adaptation any feature which makes an animal or plant adapted for a particular environment and way of life

aerobic respiration the oxygen-using chemical reaction from which most living things obtain their energy

aim the purpose of an experiment

algae a group of primitive plants

altitude height above sea level

alveoli (singular: alveolus) tiny air sacs in mammal lungs, providing a large surface area, where gas exchange takes place

amplitude the distance between the middle position of a vibrating object and the furthest point that it reaches in its vibration

angle of incidence angle between the incident ray and the normal

angle of reflection angle between the reflected ray and the normal

anther male part of a flower: produces and holds the pollen

artery any large blood vessel that carries blood away from the heart

asexual reproduction the process by which a single organism reproduces itself, producing offspring that have identical DNA to their parent

asteroid a rocky body of irregular shape (smaller than the moon) that revolves around the Sun

asteroid belt a region between Mars and Jupiter in which most asteroids in our solar system are located

astronomer a scientist who studies stars, planets and other objects in the universe

atmosphere the 'blanket' of gases around a planet

atomic number the number of protons in the nucleus of the atom of an element. This is different for each element

atomic mass a measure of how massive an atom is, compared to hydrogen atoms

atrium (plural: atria) one of the two upper chambers of the heart. These collect blood that is returning from the body or the lungs

audible able to be heard

audible sound sound that can be heard by humans

axon long extension to a nerve cell that relays electric impulses away from the cell body to another nerve cell

bar a unit of air pressure, equal to normal atmospheric pressure

binary fission a form of asexual reproduction in which a living cell divides in two

blood vessel tube along which blood is pumped around the body

body system set of organs that work together to carry out a particular function

boiling point (BP) the temperature at which a liquid boils, starts bubbling and turns into a gas

breathing muscular movements that move air in an out of the lungs

Bunsen burner a single gas jet used in science laboratories for heating

camouflaged having an appearance that blends in with the surroundings, making a species hard to see in its habitat. This is one way in which a species can help protect itself from predators

capillaries blood vessels with a very small diameter (approximately the diameter of a red blood cell) and walls that are only one cell thick. These are the blood vessels involved in the exchange of materials between the blood and body cells

carbohydrate a group of chemicals, including glucose and starch

carbon dioxide the gas that has molecules made from one atom of carbon and two atoms of oxygen. It is a greenhouse gas

carpel the female part of a flower. It consists of the ovules, ovary, style and stigma

cartilage firm but flexible body tissue, different to bone

cell membrane very thin outer layer of a plant or animal cell, controls the movement of substances in and out of the cell

cell wall thick outermost layer of plant cells

cells the tiny building blocks of most living things

cell (cellular) respiration the chemical reaction in body cells which provides energy

cellulose the complex carbohydrate in plants that makes up their wood, bark, cell walls and fibres

central nervous system the brain and spinal cord

centrifuge a machine that can make a suspension settle quickly by spinning it very fast

cervix the ring of muscles at the entrance to the uterus

changing state when a particular substance changes from one state (solid, liquid or gas) to another as a result of heating or cooling

characteristics features, including how an organism is built, how it behaves or how it functions internally

chemical bonds the forces of attraction between atoms or ions

chemical formula a shorthand way of showing which elements are present in a substance, and the relative numbers of its atoms

chemical reactivity the ability to react with other substances

chlorophyll green substance found in plants: absorbs the light energy needed for photosynthesis

chloroplast chlorophyll-containing organelle found in some plant cells

circulatory system the system that transports nutrients and oxygen to all the cells in the body and transports wastes away from the cells. It consists of blood, blood vessels and heart

classes sets of similar organisms in a phylum

classify to sort things into similar groups

coal-fired power stations power stations in which coal is burnt to boil water and produce the steam that turns turbines

cochlea the fluid-filled organ shaped like a snail shell in which sound waves are converted to electrical signals

colloid a cloudy mixture that cannot be separated by standing or filtering

colour blindness a condition in which the cone system on the retina of the eye is faulty, making it difficult to distinguish between certain colours

coloured filter a transparent material that will only transmit certain colours and reflect the rest

column graph a graph in which the change in a quantity is shown by drawing a series of columns

combustion burning

complementary colours two or more coloured lights that can be projected onto a white surface and together will produce white light

compound a chemical substance made from two or more elements combined together

compressible able to be pressed or squashed into a smaller volume

compression the part of a wave in which the particles are pressed closer together than they were before the sound wave reached them

concave lens a lens that is thicker at the edges than in the centre

concave mirror curved mirror that has its reflective surface on the inside of the curve

concentration (of a solution) a measure of how much of a substance is dissolved in each litre of a solution

conclusion what the results suggest or tell you

condensed changed from a gas into a liquid

conduction (of heat) the transfer of heat from a hot material to a less hot material but only if they are in contact

conductivity (heat) a measure of how easy it is for heat energy to flow through a material

control the version of an experiment that does not include the factor you are testing, in order to provide a basis for comparison

convection current a mass movement of liquids and gases, as a result of heating and/or cooling

convex lens a lens that is thicker in the middle than at the edges

convex mirror curved mirror that has its reflective surface on the outside of the curve

cornea the transparent curved front surface of the eye

corona the right glowing outer layer of the Sun

coverslip very thin piece of glass that is used to cover a specimen on a microscope slide

cross-pollination when the pollen from one plant is transferred to the carpel of another plant of the same species

cross-section a slice through a shape

crust the solid rocky outer layer of Earth

cyan blue-green light, the whole spectrum with red removed

decanting pouring off a liquid that has solids or denser liquids settled at the bottom, leaving them behind

decibel (dB) unit used for measuring sound intensity

deduction the process of drawing a conclusion from observations and measurements

dendrites thread-like extensions to a nerve cell that relay electrical impulses from another neuron to the cell body

density the number of grams per cubic centimetre of a material

deoxygenated blood blood that has delivered most of the oxygen it is carrying to body cells

desalination removal of salt from water

diameter the width of a sphere through its centre

diffusion the movement of particles from a region of high concentration to a region of lower concentration

digestion breakdown of big molecules into smaller ones that can be absorbed through cell membranes

digestive system the series of tubes and organs that digest food, absorb nutrients into our bloodstream, and remove wastes

dispersal to spread in different directions

dissolves mixes into water, or another solvent, forming a solution

distillation a process of removing a liquid from a mixture by evaporation and then condensing it to reform the pure liquid

diverging rays light rays that move further apart but can be traced back to an imaginary point

DNA a giant thread-like double-helix molecule that carries genetic information

dry ice solid carbon dioxide

ear canal (auditory canal) the tube along which sound travels from the pinna to reach the eardrum

eardrum the membrane across the ear canal that vibrates when sound waves reach it

echo sound that has reflected off a surface

eggs female gametes, also called ova

electric circuit a complete path around which electricity flows

electromagnetic radiation a form of energy that includes ultraviolet, infrared, X-rays, light and radio waves. All travel at 300,000 km per second

electrons very tiny negatively charged particles that spin around the nucleus of an atom

ellipse shape of Earth's orbit around the Sun

empirical any knowledge based on experience, and evidence

embryo term given to the developing child when its cells begin to specialise and some organs start to form

endangered species a species that has a high risk of extinction in the near future testing

energy transformation the changing of energy from one form to another, such as from electric energy to light energy

engineers scientists who use their knowledge of science and mathematics to make or build things

enhanced greenhouse effect the additional warming of Earth's surface that is caused by the production of more carbon dioxide than can be used naturally

Eustachian tubes tubes between the nose and middle ears

evaporated (vapourised) changed from a liquid into a gas (vapour)

evidence facts or science showing that something exists or is true

explosion a violent rush of gas outwards when pressure inside is higher than the outside pressure

fair test a test in which all variables are kept the same except for one

family a set of similar organisms in an order

fermentation process in which sugars are converted to alcohol and carbon dioxide by organisms such as yeast organisms to obtain their energy

fertile offspring young that can have offspring of their own when they reach maturity

fertilisation (conception) joining of male and female sex cells, producing a new single cell that contains both of their DNA

filter funnel a special piece of equipment made for filtering liquids in the laboratory

filter paper special paper designed for filtering solids out of water or other liquids

filtering using a filter paper or filter to remove solid particles from water or other liquids

filtrate the liquid that passes through the filter paper during a filtration

flammable (inflammable) easily able to catch on fire

flowing movement of a liquid or gas from place to place

focal point the point at which convergent rays meet

focusing adjusting the microscope to obtain a clear image of the object

foetus (fetus) the name given to the developing child after about 6 to 8 weeks. Before this: embryo

fossil fuel any burnable substance made over millions of years from dead plants and animals

frequency (of vibration) number of times a vibrating object completes one back-and-forth movement each second. This scientific quantity is related to the pitch of a sound

freezing point the temperature at which a liquid changes into a solid. This is the same as the melting point

fuel a substance that can react with oxygen, burning with a flame

gametes sex cells. These only contain half the DNA present in other body cells

gas exchange the diffusion of oxygen in one direction and of carbon dioxide in the other direction

gas giants large planets made mostly made of gases

geologist a scientist who studies Earth

geothermal power station a power station that uses the heat of underground rocks to produce the steam to make electricity

global warming the enhanced greenhouse effect, in which an increased amount of gases like carbon dioxide trap extra heat by preventing it from being radiated into space, and in this way raise the temperature of the planet

glucose simple sugar that is used for cellular respiration. Produced during photosynthesis

gravity an invisible force of attraction between all things that have mass

greenhouse gas a gas that warms Earth's surface

group a vertical column of elements in the periodic table. Elements in the same group have similar chemical properties

haemoglobin (hemoglobin) protein molecule present in red blood cells that transports oxygen to the cells

hammer, anvil and stirrup the three smallest bones in the human body are located in the middle ear and transmit and amplify sound; also known as ossicles

heart the muscular pump that pumps blood around the blood vessels

heat energy a form of energy related to the speed of particle movement

heat transfer the flow of heat from a warmer object to a cooler object

hectopascal (hPa) a measurement of air pressure for weather forecasting

hertz (Hz) unit of sound frequency

high concentration more particles are packed together into a given space

hydrosphere the layer of water that covers most of Earth, in sea, lakes and ice

hypothesis an idea which could explain a set of observations. A hypothesis may contain the words 'If ..., then ...'

ignite burst into flame

image picture of an object formed by a mirror or other reflective surface

implosion a violent rush of gas inwards when pressure outside is higher than inside

incident ray incoming light ray onto a surface

indicator substance that changes colour depending on the pH of the solution to which it is added

inert non-reactive under normal conditions

infertile unable to reproduce

informed opinion an opinion based on knowledge

interior planets planets found between Earth and the Sun (Mercury and Venus)

interstellar space term astronomers use to describe all space beyond the solar system — the space between the stars

inventions things that have been designed to do something useful, often to solve a problem

ions atoms with an electrical charge that has resulted from the loss or gain of one or more outer electrons

iris the coloured muscular tissue that controls the size of the pupil

isotopes atoms of the same element that have different numbers of neutrons in their nuclei

kaitiakitanga the Maori principle of guardianship, and of caring for resources

kaumatua Maori elder, or senior person of knowledge

kilojoule (kJ) unit of heat energy

kilopascal (kPa) a metric measurement for gas pressure in containers and pipes

kinetic energy energy possessed by a moving object

kinetic theory a scientific theory that the behaviour of solids, liquids and gases is due to the way their particles vibrate and join with each other

kingdoms the largest groupings of living things

laboratories (labs) rooms where scientific experiments are carried out

law of conservation of energy the scientific rule that energy cannot be made or destroyed, although it can change form

law of reflection mathematical relationship between the angle of incidence and angle of reflection. These angles are equal for a given light ray when it hits a plane or curved surface

lens a curved transparent object

ISBN: 9780170214650

lens (of an eye) the clear elastic tissue that focuses light in the eye. Its shape can change to focus on closer and more distant objects

ligament band of tough, slightly elastic fibre that holds bones together at a joint

light ray thin pencil-like beam of light

light year distance travelled by light in one year, used to measure the vast distances of space

limewater (Ca(OH)$_2$) a solution that turns milky when carbon dioxide bubbles through it

line graph a graph in which the plotted points are joined with a line or curve

lithosphere the layer of Earth that includes the crust and the outer rocky mantle

logarithmic a scale of measurement based on powers of 10

low concentration few particles in a given volume

luminous produces light

lunar eclipse event when Earth passes between the Sun and the moon, catching the moon in its shadow

lung capacity an estimate of how much air is in the lungs

mammal warm-blooded animals, generally have fur or hair, give birth to live young and feed young milk

mantle the layer of the earth underneath the crust, 3000 km thick

marsupial mammal that gives birth to immature young that are then sheltered in a pouch

Matariki Maori word for a group of stars also known as the Pleiades

melting point (MP) the temperature at which a solid changes into a liquid

metals chemical elements that are shiny, good conductors of electricity, and tend to give away electrons when they react

methane a simple carbon compound, part of natural gas

microscope slide thin rectangle of glass on which small specimens are placed for examination under a microscope

microscopic can only be seen using a microscope

middle ear the air-filled space spanned by the hammer, anvil and stirrup

miscible able to be completely mixed

mixture a substance that can be separated into two or more different compounds and/or elements by a separation method such as distilling or filtering

molecule a group of atoms joined together

monocular microscope microscope that has a single eyepiece for viewing the object with one eye

moon (natural satellite) a planet-like body that revolves around a planet (which is larger than it)

multicellular organism organism that is built from more than one cell

neap tide extra-low tide created when the Sun and moon pull on the oceans at right angles to one another

near space a term that astronomers use to describe the solar system

nerve bundle of nerve fibres

nerve cells cells that are linked together and send messages from one place in the body to another

nerve fibres another name for dendrites and axons

nervous system the system of brain, spinal cord and nerves that communicates with body cells and controls (regulates) what happens in the body by means of electrical impulses

neuron a nerve cell, which is a specialised cell that can transmit an electrical impulse (also spelled neurone)

neutrons tiny uncharged particles found in the nucleus of an atom. They are of similar size and mass to protons

night the time of darkness for the part of a planet that is facing away from the Sun

nitrogen gas making up 78% of the air. It does not burn or react easily

nitrogen-fixation changing nitrogen into forms that plants can use to produce amino acids and hence proteins

non-compressible not able to be pressed into a smaller space

normal line a line drawn at 90° (right angles) to the surface of a mirror or lens

observation something you see, feel, hear, smell or taste (although tasting is rarely used in scientific observation because it can be dangerous)

one-way valve valve that only allows fluid to flow through it in one direction

Oort cloud vast cloud of rocks and ice on the boundary of the solar system where comets are born

opaque does not let light through

optic nerve transmits visual images to the brain as electrical signals

order a set of similar organisms in a class

organ set of tissues that carries out a particular function

organelle part of a cell

organism a living thing

osmosis a special case of diffusion which occurs when two different solutions are separated by a permeable membrane. Water moves from the side in which there is more water, to the side in which there is less water

ossicles the three small bones of the ear: hammer, anvil and stirrup

outer ear the ear canal and pinna

oval window membrane across the entry to the cochlea, to which the stirrup is attached

ovary the organ in which eggs are produced and kept

ovum (egg) female gamete

oxygenated blood blood in which the haemoglobin molecules are carrying their load of oxygen. This blood is bright red

ozone a form of oxygen that above the stratosphere absorbs UV radiation from the Sun, but at ground level is a pollutant

ozone layer the layer of air in the upper atmosphere that contains ozone and shields us from most UV radiation from the Sun

paper chromatography technique for separating components in a solution by allowing it to soak up special absorbent paper (chromatography paper). The components travel up at different speeds and hence become separated

parallax error the error that arises when you read a scale at an angle instead of directly in front of it

partial solar eclipse event in which the Sun is partially obscured by the moon as the moon passes almost directly between Earth and the Sun

particle theory a scientific theory stating that every substance consists of tiny particles in motion

pathogens disease-causing organisms

Periodic table a table of the elements listed in order of increasing atomic number, and elements with similar properties placed in the same vertical column

pH scale scale for measuring the acidity of a solution. The more acidic the solution, the lower the number of the scale. pH scale is from 0 to 14

phases of the moon different shapes that the moon appears to have

phloem plant tissue with tubular cells for carrying sugar solutions

photosynthesis chemical reaction in green plants, in which the plant uses light energy to convert carbon dioxide and water into both glucose and oxygen.

phylum a set of similar organisms in a kingdom

physical change a reversible change, such as a change of state or shape

physicist a scientist who studies some aspect of the physical world such as movement, light, electricity or sound

pinna outside 'shell' of the ear that funnels sound waves

plane mirror flat mirror that produces an image the same distance behind the mirror as the object is in front

plant kingdom the set of organisms that produce their own glucose by photosynthesis

pollen tiny grains released by plants, and containing the male sex cells

pollen tube a tube that begins to grow from a pollen grain down into the stigma towards the ovary

pollination the movement of pollen from one flower to another, by wind or animals

positive feedback a sequence of responses that leads to further increase. Also known as a 'vicious circle'

precipitation reaction a chemical reaction in a solution, in which an insoluble solid is produced

prediction a statement about what you think is going to happen, based on reasons

pressure a measure of the force exerted on a certain area of surface

primary additive colours the three colours of light that can produce white light when mixed equally: red, blue and green

procedure steps to follow to perform an experiment

product a substance formed in a chemical reaction

properties special features of a particular substance or living thing

protein giant molecules in all living things. Contain nitrogen atoms

protons positively charged particle in the nucleus of an atom

pulmonary circulation circulation of the blood from the heart to the lungs and back to the heart

pulse the ripple-like movement in an artery wall cause by the surge of pressure of blood inside it. This can be used to determine how may times the heart beats each minute

pupil the dark opening in the centre of the iris, allowing light to enter

ISBN: 9780170214650

radiation the transfer of heat or light and other forms of electromagnetic energy through space, at the speed of light

rarefaction the part of a sound wave in which the particles are further apart than they were before the sound wave reached them

reactant substance that reacts with other substances and thus undergoes a chemical change

real image the image of an object that can be seen on a screen

red blood cells disc-shaped cells found in blood. Each one contains millions of haemoglobin molecules. Responsible for delivering oxygen to body cells

reflex an automatic response to a stimulus

reflex arc the pathway of nerve cells involved in a reflex action

refracted ray light ray bent from its original path when it enter a transparent material

reliable results results that will be obtained if others repeat the experiment

renewable energy energy made by resources that will never run out, like the Sun or wind

reproduction making more members of the same species

residue the substance remaining in a filter paper after filtration

respiration the chemical process of releasing energy from foods such as glucose

respiratory surface any part of an animal where gas exchange occurs

result what you observe or measure

retina a light-sensitive surface inside the eye

reverberation continuing echoes of a sound

salt compound with metal and non-metal elements combined

saturated solution a solution that contains the maximum amount of solute dissolved in the solvent at that temperature

scientific method the process of observing things, trying to explain them, conducting experiments to test these ideas, and modifying the ideas if something unexpected happens

secondary (or subtractive) colour colours produced by a mix of two primary colours equally. The three secondary colours are cyan (blue and green), magenta (blue and red) and yellow (green and red)

self-pollination when pollination occurs within an individual plant

semi-circular canals (organ of balance) the three fluid-filled loops located in the inner ear that control balance

sensory neurons neurons that carry messages from the body to the central nervous system

sensory receptors parts of highly specialised nerve cells that are capable of detecting information and relaying it by electrical impulses. Example: rods and cones

separating funnel a funnel that has a tap at the bottom and is used to separate immiscible liquids

septum muscular wall that separates the right side of the heart from the left side

sex cells eggs and sperm (also collectively known as gametes)

SI Systeme Internationale, which uses units such as joules, watts and kilograms but not horsepower or miles or calories

single-celled (unicellular) organism organism made up from only one cell

solar eclipse event when the Sun is obscured by the moon as the moon passes directly between Earth and the Sun

solar system the Sun and all the bodies that revolve around it, such as planets and asteroids

solidify change from a liquid to a solid when cooled

solubility a measure of the amount of a substance dissolved in each 100 g of solvent

solute the dissolved substance in a solution

solution a mixture in which the particles of the different substances cannot be seen, and cannot be separated by filtration alone

solvent a liquid that dissolves a solute

sonar technology based on echolocation

specialised cells cells in multicellular organisms that do particular jobs

species a group of similar organisms that are able to breed with each other and produce living, fertile young

spectrum the band of colours seen when light is broken up by a prism

speed of sound the distance a sound wave travels out from its source each second

specialised adapted, modified or equipped for a particular purpose. The opposite of generalised

sperm male gamete

spinal cord the main information pathway between the brain and the rest of the body. Housed inside the bones of the spinal column

spring tide extra-high tide created when the moon and Sun both pull on the oceans from the same direction

stamen male part of the flower. It consists of an anther and a filament

starch the complex carbohydrate made by plants that is mainly stored in their seeds

stereo-microscope a microscope that produces a three-dimensional image of whole objects

stigma the pollen-receiving female part of a flower

stimulus whatever activates a sensory cell in the nervous system or causes a response in the endocrine system

stomata the holes in the leaves of plants through which they absorb carbon dioxide for photosynthesis and release the oxygen produced in photosynthesis. Stomata also lose water through transpiration

sublimation a change directly from a solid to a gas or vice versa

substance something that takes up space and has weight

sugars small-molecule carbohydrates

supersaturated solution a solution made by cooling a saturated solution slowly so that no crystals form

suspension a mixture in which the substance stays suspended in the water for a little while then sinks to the bottom

sustainable able to be continued, for example using resources so that they will not run out

synapse small gap between two nerve cells

tendon tough white fibre that attaches muscles to bone

terrestrial planets smaller, rocky Earth-like planets

theory a major idea that tries to explain a large set of observations

thermometer an instrument used to measure temperature

tissue grouping of similar cells

trachea windpipe taking air to and from the lungs

translucent lets some light through. Objects cannot be clearly seen through a translucent material

transparent lets most light through. Objects can be clearly seen through a transparent material

transpiration loss of water vapour by plants, mostly through their leaves

ultrasound high frequency sound that cannot be heard by humans. Above 20 kHz

umbilical cord the artery and vein that connects the embryo or foetus to the placenta along which blood flows, carrying nutrients and oxygen to and wastes away from the foetus

vacuum empty space with no substance in it, or almost none

valve device that allows one-way flow of blood (or any fluid)

variables factors in an experiment that may affect the results

vein blood vessel that carries blood back to the heart

ventricle The two lower chambers of the heart. The right ventricle pump blood to the lungs, the left to the body

vertebrae (singular: vertebra) the hard bones of the spinal column

vibrating moving back-and-forth

viscous sticky and 'gluey'; very slow flowing

volume the space that something takes up

warm-blooded produces its own body heat

waste products substances that living things make in their bodies that they cannot use and will poison them if they build up too much

watt the unit of power

wave motion movement that comes in patterns, for example, sound waves travel from the sound source as a series of compressions and rarefactions

wavelength the distance between two successive wave peaks

word equation a word summary to show the reactants and products of a chemical reaction

xylem long tube-like plant cells that carry water upwards

year the time a planet takes to revolve around the Sun

yeasts microscopic organisms that ferment fruits and grains to obtain their energy

ISBN: 9780170214650

Periodic table of the elements

KEY

atomic number	26
element symbol	Fe
element name	iron

Key colours: metal, metalloid, non-metal

Group	1	2	3	4	5	6	7	8	9	10	11	12	13	14	15	16	17	18
Period 1	1 H hydrogen																	2 He helium
Period 2	3 Li lithium	4 Be beryllium											5 B boron	6 C carbon	7 N nitrogen	8 O oxygen	9 F fluorine	10 Ne neon
Period 3	11 Na sodium	12 Mg magnesium											13 Al aluminium	14 Si silicon	15 P phosphorus	16 S sulfur	17 Cl chlorine	18 Ar argon
Period 4	19 K potassium	20 Ca calcium	21 Sc scandium	22 Ti titanium	23 V vanadium	24 Cr chromium	25 Mn manganese	26 Fe iron	27 Co cobalt	28 Ni nickel	29 Cu copper	30 Zn zinc	31 Ga gallium	32 Ge germanium	33 As arsenic	34 Se selenium	35 Br bromine	36 Kr krypton
Period 5	37 Rb rubidium	38 Sr strontium	39 Y yttrium	40 Zr zirconium	41 Nb niobium	42 Mo molybdenum	43 Tc technetium	44 Ru ruthenium	45 Rh rhodium	46 Pd palladium	47 Ag silver	48 Cd cadmium	49 In indium	50 Sn tin	51 Sb antimony	52 Te tellurium	53 I iodine	54 Xe xenon
Period 6	55 Cs caesium	56 Ba barium	57 La lanthanum	72 Hf hafnium	73 Ta tantalum	74 W tungsten	75 Re rhenium	76 Os osmium	77 Ir iridium	78 Pt platinum	79 Au gold	80 Hg mercury	81 Tl thallium	82 Pb lead	83 Bi bismuth	84 Po polonium	85 At astatine	86 Rn radon
Period 7	87 Fr francium	88 Ra radium	89 Ac actinium	104 Rf rutherfordium	105 Db dubnium	106 Sg seaborgium	107 Bh bohrium	108 Hs hassium	109 Mt meitnerium	110 Ds darmstadtium	111 Rg roentgenium							

Period 6 (lanthanides):

58 Ce cerium	59 Pr praseodymium	60 Nd neodymium	61 Pm promethium	62 Sm samarium	63 Eu europium	64 Gd gadolinium	65 Tb terbium	66 Dy dysprosium	67 Ho holmium	68 Er erbium	69 Tm thulium	70 Yb ytterbium	71 Lu lutetium

Period 7 (actinides):

90 Th thorium	91 Pa protactinium	92 U uranium	93 Np neptunium	94 Pu plutonium	95 Am americium	96 Cm curium	97 Bk berkelium	98 Cf californium	99 Es einsteinium	100 Fm fermium	101 Md mendelevium	102 No nobelium	103 Lr lawrencium

ISBN: 9780170214650